SpringerBriefs in Modern Perspectives on Disability Research

Series Editors

Gabriel Bennett, Independent Researcher, Klemzig, Australia

Emma Goodall, Healthy Possibilities, Seaford, Australia

This book series on disability research is a comprehensive collection of research on disability and related issues. The series is designed to promote interdisciplinary collaboration and exchange, bringing together scholars and practitioners from different fields to share their perspectives and insights. Disability research is an interdisciplinary field that examines the social, cultural, historical, and political dimensions of disability. It encompasses a wide range of topics, including disability rights, accessibility, assistive technologies, healthcare, education, employment, and social welfare. Disability research scholars employ a range of theoretical and methodological approaches to understand the experiences of people with disabilities, as well as the ways in which disability intersects with other social identities such as race, gender, sexuality, and class.

The series seeks to advance knowledge and understanding of disability by publishing rigorous, innovative, and relevant research. It aims to promote disability rights and social justice by highlighting the ways in which people with disabilities are marginalized and discriminated against in society, and advocating for greater social inclusion and accessibility. The series also seeks to inform policy and practice by disseminating research findings that can help to shape policy decisions and contribute to positive social change.

Arshad Nawaz Malik · Huma Riaz · Suman Sheraz

Emerging Technologies in the Rehabilitation of Physical Disabilities

 Springer

Arshad Nawaz Malik
Faculty of Rehabilitation and Allied Health
Sciences
Riphah International University
Islamabad, Pakistan

Huma Riaz
Faculty of Rehabilitation and Allied Health
Sciences
Riphah International University
Islamabad, Pakistan

Suman Sheraz
Faculty of Rehabilitation and Allied Health
Sciences
Riphah International University
Islamabad, Pakistan

ISSN 3004-9709   ISSN 3004-9717 (electronic)
SpringerBriefs in Modern Perspectives on Disability Research
ISBN 978-981-92-1342-9   ISBN 978-981-92-1343-6 (eBook)
https://doi.org/10.1007/978-981-92-1343-6

This Springer imprint is published by the registered company Springer Nature Singapore Pte Ltd.
The registered company address is: 152 Beach Road, #21-01/04 Gateway East, Singapore 189721,
Singapore

If disposing of this product, please recycle the paper.

This book is dedicated to persons with disabilities whose resilience and determination continue to inspire progress in rehabilitation, and to the students and future rehabilitation professionals whose curiosity, commitment, and passion will shape the future of healthcare.

May this work contribute to improving lives and advancing the field of disability rehabilitation.

Preface

Advancements in technology are transforming the landscape of healthcare, and rehabilitation sciences are no exception. Over the past decade, emerging technologies such as Robotics, Virtual and Augmented Reality, Wearable Sensors, Artificial Intelligence, Telerehabilitation, and Digital Health Platforms have begun to reshape the way physical disabilities are assessed, monitored, and managed. These innovations offer new opportunities to enhance patient outcomes, improve accessibility to rehabilitation services, and support clinicians in delivering more personalised and evidence-informed care.

The idea for this book emerged from the growing recognition that rehabilitation professionals must not only understand traditional therapeutic approaches but also remain informed about technological developments that are increasingly influencing clinical practice and research. Despite the rapid expansion of these technologies, resources that present their applications within the context of rehabilitation in a structured and accessible manner remain limited. This book was, therefore, developed to bridge the gap by bringing together key concepts, practical applications, and current developments in rehabilitation technologies.

This book is envisioned for rehabilitation professionals, including physiatrists, physiotherapists, occupational therapists, and other healthcare providers involved in the management of physical disabilities, as well as biomedical engineers, technology developers, and software and AI specialists interested in designing rehabilitation-focused technologies. It will also be valuable to faculty members, researchers, students, healthcare professionals, policymakers, and advocates working toward innovation, accessibility, and inclusive rehabilitation for people with disabilities.

The authors would like to acknowledge the collective efforts of researchers, clinicians, and innovators whose work continues to advance the field of rehabilitation and improve the quality of life for individuals with physical disabilities.

Islamabad, Pakistan

Arshad Nawaz Malik
Huma Riaz
Suman Sheraz

Competing Interests The authors have no competing interests to declare that are relevant to the content of this manuscript.

Contents

1 Technology Integrated Disability Rehabilitation: A Modern Paradigm Shift 1
Arshad Nawaz Malik, Huma Riaz, and Suman Sheraz
1.1 Introduction 2
1.2 Disability Rehabilitation 2
1.3 Limitations of Traditional Rehabilitation Approaches 3
1.4 Paradigm Shift from Traditional to Technology-Integrated Rehabilitation 4
1.5 Spectrum of Technologies in Modern Rehabilitation 5
1.6 Technology Integration in Physical Disability Rehabilitation 11
1.7 Conclusion 13
References 14

2 Emerging Technologies in Neurological Rehabilitation 17
Arshad Nawaz Malik
2.1 Global Burden and Classification of Neurological Disorders 18
2.2 Impact of Neurological Impairments and Functional Limitations 19
2.3 Mechanism of Motor Recovery in Neurological Disorders 20
2.4 Limitations of Conventional Neurorehabilitation Approaches 21
2.5 Rationale for Emerging Technologies in Neurorehabilitation 22
2.6 Emerging Technologies in Neurorehabilitation 23
2.7 Disease-Specific Applications of Emerging Technologies 26
2.8 Limitations and Challenges of Emerging Technologies 38
2.9 Conclusion 40
References 40

3 Technological Innovations in Rehabilitation of Musculoskeletal Disabilities 47

Huma Riaz and Suman Sheraz

3.1 Overview of Musculoskeletal Disorders and Disabilities 48

3.2 Global Burden of MSDs and Related Disability 48

3.3 Musculoskeletal Disability within the International Classification of Functioning (ICF) Framework 50

3.4 Impairments and Functional Limitations of Common MSDs 51

3.5 Mechanisms of Recovery in Musculoskeletal Rehabilitation 53

3.6 Technological Innovations in Musculoskeletal Assessment 55

3.7 Technological Innovations in Musculoskeletal Management 59

3.8 Future Directions in MSK Technological Rehabilitation 66

References 66

4 Technological Interventions for Paediatric Motor Disabilities 73

Arshad Nawaz Malik and Suman Sheraz

4.1 Overview of Developmental and Paediatric Disabilities 73

4.2 Mechanisms of Recovery in the Developing Brain 76

4.3 Limitations of Conventional Paediatric Rehabilitation 78

4.4 Rationale for Integrating Emerging Technologies 79

4.5 Emerging Technological Interventions in Paediatric Rehabilitation 79

4.6 Disease-Specific Applications of Technological Interventions 83

4.7 Challenges and Future Directions 90

References 91

5 Geriatric Physical Disabilities and Technological Advancements 95

Suman Sheraz

5.1 Physical Disability, Frailty, and Function in Older Adults 95

5.2 Geriatric Rehabilitation: Overview, Limitations, and the Emerging Role of Technology 99

5.3 Emerging Technologies in Geriatric Rehabilitation 102

5.4 Technology-Enhanced Geriatric Rehabilitation Protocols 111

5.5 Challenges in Technology-Enabled Geriatric Rehabilitation 114

5.6 Conclusion and Future Directions 116

References 116

6 Women's Health Disabilities and Technology-Enabled Rehabilitation 121

Huma Riaz

6.1 Introduction 122

6.2 Global Burden and Prevalence of Disability Conditions in Women 122

6.3 Pathophysiological Mechanisms of Disability Conditions 127
6.4 Emerging Technologies in the Rehabilitation
     of Postmenopausal Osteoporosis and Fragility
     Fractures 130
6.5 Emerging Technologies in the Rehabilitation of Breast
     Cancer-Related Lymphedema Disability 136
6.6 Emerging Technologies in the Management
     of Post-hysterectomy Chronic Pelvic Pain 140
6.7 Conclusion and Future Directions 144
References 145

**7 Challenges and Future Directions of Emerging Technologies
   in Disability Rehabilitation** 153
   Arshad Nawaz Malik and Huma Riaz
7.1 Introduction 154
7.2 Key Challenges in Technology-Driven Rehabilitation 154
7.3 Future Directions in Rehabilitation Technologies 157
7.4 Integrating Emerging Technologies into Sustainable
     Rehabilitation Models 159
7.5 Conclusion and Future Direction 162
References 162

About the Authors

Prof. Dr. Arshad Nawaz Malik is a physical therapist, researcher, and academic leader with nearly two decades of experience in neurorehabilitation, focusing on stroke recovery and disability management. He holds a PhD in Stroke Rehabilitation and is a certified Neurorehabilitation Therapist from Seoul National University, South Korea. He currently serves as Associate Dean of the Faculty of Rehabilitation and Allied Health Sciences at Riphah International University, Islamabad, Pakistan. His academic and clinical work emphasises the integration of innovative technologies into neurorehabilitation practice.

Dr. Malik has authored numerous publications in peer-reviewed journals and serves as an editor for the *Journal of Riphah College of Rehabilitation Sciences*. He supervises postgraduate and doctoral research projects and collaborates internationally on multidisciplinary studies in neurorehabilitation and disability sciences. His work also contributes to shaping national-level rehabilitation policy and professional development frameworks. He has published extensively on topics such as virtual reality, task-oriented approaches to stroke mobility recovery, balance dysfunction, and rehabilitation research. Dr. Malik's leadership extends to organising national and international neurorehabilitation events, including symposia on stroke rehabilitation, and he continues to shape the development and delivery of advanced clinical education and practice in rehabilitation sciences.

Prof. Dr. Huma Riaz is a physical therapist, academic leader, and researcher with over 18 years of experience in rehabilitation sciences. She holds a PhD in Rehabilitation Sciences with a research focus on postmenopausal osteoporosis and currently serves as the Principal of Riphah College of Rehabilitation and Allied Health Sciences, Riphah International University, Islamabad, Pakistan. Her academic and clinical work focuses on women's health, musculoskeletal rehabilitation, and the integration of technological innovations in physical therapy.

Dr. Riaz holds additional qualifications in medical education and healthcare leadership and has played a key role in curriculum development and academic program advancement. She has supervised numerous postgraduate research projects and has secured several national research grants, including a competitive grant for telere-habilitation in gestational diabetes. Dr. Riaz has authored multiple peer-reviewed publications in indexed journals and serves as Associate Editor of the *Journal of Riphah College of Rehabilitation Sciences*. She has received national recognition for her contributions to rehabilitation education and research and actively collaborates with national and international partners. Her current research interests include digital health, women's bone health, pelvic rehabilitation, and personalised rehabilitation interventions for physical disabilities.

Dr. Suman Sheraz is a physical therapist, researcher, and academician with over 13 years of clinical and academic experience in rehabilitation sciences and physical therapy education. She holds a PhD in Rehabilitation Sciences and currently serves as an Associate Professor and Head of the Department of Physical Therapy at Riphah International University, Islamabad, Pakistan. Her research focuses on rehabilitation strategies for individuals with chronic conditions and disabilities, with a particular interest in diabetes management, functional health in older adults, community-based rehabilitation, and promoting physical activity to improve functional independence and quality of life.

Dr. Sheraz has authored numerous publications in peer-reviewed journals and has presented her work at national and international conferences. She collaborates with researchers from the United Kingdom, Italy, Spain, and Canada and contributes to curriculum development, postgraduate clinical training, and initiatives to strengthen rehabilitation education and research capacity. She also serves as an Associate Editor of the *Journal of Riphah College of Rehabilitation Sciences* and has been recognised for her contributions to digital education and innovation in rehabilitation sciences.

Abbreviations

AAL	Ambient Assisted Living (AAL)
ADLS	Activities of Daily Living
AI	Artificial Intelligence
AR	Augmented Reality
BCI	Brain Computer Interface
BCRL	Breast Cancer-Related Lymphedema
BF	Biofeedback
BMD	Bone Mineral Density
CP	Cerebral Palsy
CPP	Chronic Pelvic Pain
DALYs	Disability-Adjusted Life Years
GBD	Global Burden of Disease
HAL	Hybrid Assistive Limb
HMDs	Head-Mounted Displays
IMU	Inertial Measurement Units
IoT	Internet of Things
LMIC	Low-Middle-Income Countries
mHealth	Mobile Health
ML	Machine Learning
MSD	Musculoskeletal Disorders
NDs	Neurological Disorders
NIBs	Non-Invasive Brain Stimulation
NLP	Natural Language Processing
PD	Physical Disability
QoL	Quality of Life
RAGT	Robot-Assisted Gait Training
RAT	Robot-Assisted Training
RMS	Remote Monitoring Systems
rTMS	repetitive Transcranial Magnetic Stimulation
SDGs	Sustainable Development Goals
SRG	Soft Robotic Gloves

tDCS	Transcranial Direct Current Stimulation
TR	Tele-Rehabilitation
VPA	Virtual Physiotherapy Assistants
VR	Virtual Reality
WBV	Whole-Body Vibration
WDs	Wearable Devices
WHO	World Health Organisation
WS	Wearable Sensors
YLD	Years Lived with Disability

Chapter 1
Technology Integrated Disability Rehabilitation: A Modern Paradigm Shift

Arshad Nawaz Malik⬥, Huma Riaz⬥, and Suman Sheraz⬥

Abstract Rehabilitation for individuals with physical disabilities is undergoing a significant transformation with the integration of innovative technologies. Conventional rehabilitation approaches have certain limitations in achieving a functional recovery and outcome. This chapter explains the shift from traditional human-dependent models to technology-based interventions, ranging from robotics and virtual reality to wearable sensors, artificial intelligence, and telerehabilitation platforms. These innovative technologies offer better recovery, personalisation, and engagement, enhancing functional recovery in patients with physical disabilities. The chapter highlights the integration of technology-enabled rehabilitation for condition-specific disabilities, including neurological, musculoskeletal, paediatric, geriatric, and women's health conditions. The chapter sets the stage for subsequent chapters that explore technological rehabilitation solutions for physical disabilities across a range of conditions, populations, and health systems.

Keywords Digital health · Disability rehabilitation · Emerging technology · Persons with disabilities · Physical disability

A. N. Malik (✉) · H. Riaz · S. Sheraz
Faculty of Rehabilitation and Allied Health Sciences, Riphah International University, Islamabad, Pakistan
e-mail: arshad.nawaz@riphah.edu.pk

H. Riaz
e-mail: huma.riaz@riphah.edu.pk

S. Sheraz
e-mail: suman.sheraz@riphah.edu.pk

© The Author(s), under exclusive license to Springer Nature Singapore Pte Ltd. 2026
A. N. Malik et al., *Emerging Technologies in the Rehabilitation of Physical Disabilities*, SpringerBriefs in Modern Perspectives on Disability Research,
https://doi.org/10.1007/978-981-92-1343-6_1

1.1 Introduction

Physical disability (PD) remains a major global health challenge, affecting approximately one billion individuals around the world (Kołłątaj et al., 2025). According to WHO, 15% of the total world population lives with some disability, and 80% of those live in low- and middle-income countries (LMICs) (Naicker et al., 2019a). PDs comprise a broad spectrum of impairments and functional limitations, ranging from cognitive involvement to mobility dysfunction. Impairments such as balance dysfunction, incoordination, weakness, spasticity, sensory loss, pain, and lack of motor control reduce the self-care capabilities, dressing, bathing, feeding, and toileting. These impairments significantly impact functional status, activity levels, and active participation in society. Functional impairments further lead to limited walking ability, independence in instrumental activities of daily living, and performance at the occupational and community levels (Cieza et al., 2020a). Limited interaction and isolation among individuals with PD increase dependence on caregivers and ultimately lead to a decline in the quality of life. Secondary complications also pose serious threats to people with PD, including fatigue, pain, increased risk of falls, deconditioning, and depression. The longstanding nature of PD can have long-term impacts on social integration, vocational engagement, psychological well-being, and overall life satisfaction. Eventually, PD becomes not only a biomedical disease, but rather it becomes a multidimensional condition that affects the physical, emotional, and social well-being of people with PD (Mennella et al., 2023a).

1.2 Disability Rehabilitation

Rehabilitation is a crucial part of multidisciplinary care for people with physical disabilities, restoring function and enabling them to perform occupational activities. The Global Burden of Diseases study (2019) reports that approximately 2.41 billion people worldwide, or one in every three individuals, will develop a health condition requiring rehabilitation services at some point in their lives. Since 1990, the rise has been approximately 63%, driven by an ageing population and a high incidence of non-communicable diseases (Cieza et al., 2020b). Rehabilitation plays a central role in restoring function, enhancing independence, and improving quality of life for patients with PDs. According to the WHO Rehabilitation 2030 initiative, rehabilitation is a fundamental health intervention deserved by Persons with Disabilities (PWD) (Naicker et al., 2019a). It is established that rehabilitation services are not considered specialised services but an essential component of healthcare systems worldwide, addressing the needs of large numbers of people. The current evidence clearly shows that rehabilitation has significantly improved function, promoted independence, encouraged participation, and enhanced quality of life in people with diverse health conditions.

The most prevalent conditions requiring rehabilitation services include musculoskeletal (low back pain, osteoarthritis), neurological conditions (Stroke, Spinal Cord Injury (SCI), Parkinson's Disease), geriatric conditions (Frailty, Fall), paediatric conditions (Cerebral Palsy, Muscular Dystrophies), and women's health conditions (osteoporosis, breast cancer-related lymphedema, and pelvic floor dysfunction). These conditions lead to physical disability, limited activity, decline in functional status, participation restriction, and declining quality of life. These conditions vary across diverse age groups, including paediatric, adult, and older adult populations, underscoring rehabilitation as a lifespan health service. Physical rehabilitation interventions can optimise function and minimise disability for those with physical impairments, but are often inaccessible to populations living in low-resource settings. The barriers include limited access to rehabilitation services, financial constraints, and a shortage of trained staff, all of which hinder the efficient and timely delivery of these services. These discrepancies significantly contribute to limited functional status, restrictions in participation, and long-term dependence on family and society. Thus, significant geographic and socioeconomic inequalities continue to limit access to rehabilitation services within low- and middle-income countries, despite the substantial demand (Kumurenzi et al., 2023).

1.3 Limitations of Traditional Rehabilitation Approaches

Traditional rehabilitation approaches are entirely therapist-dependent, physically interactive, hospital-based, and of shorter duration for effective implementation. These approaches have been considered highly effective but face multiple challenges and barriers to achieving maximum compliance. A major barrier is that workforce capacity is insufficient to meet rehabilitation needs, particularly in LMICs. There is a deficiency of staff, including physical therapists, occupational therapists, and speech therapists, in hospitals and community rehabilitation centres. As stated in the WHO's Rehabilitation 2030 Call for Action agenda, there are only about one-tenth of the required rehabilitation professionals, and only 10 skilled professionals per 10 million population. Such barriers lead to insufficient delivery of rehabilitation services to deserving populations (Gimigliano & Negrini, 2017).

There is limited access to rehabilitation services for all people with PDs, as rehabilitation services are commonly available in urban areas. People with low socioeconomic status and living in rural areas have no access to rehabilitation services. In a few LMICs, more than 50% of people with physical disabilities don't receive rehabilitation services to manage their impairments (Neill et al., 2023). Financial barriers are another factor limiting access to traditional rehabilitation services. Long-term rehabilitation care is essential in chronic health conditions with long-term disabilities, including stroke, SCI, back pain, and neurodegenerative diseases. Long-term care places a burden on the health system and is costly for families as well. These barriers and limitations of traditional rehabilitation services provide an opportunity to explore the role of emerging technologies (Naicker et al., 2019b).

1.4 Paradigm Shift from Traditional to Technology-Integrated Rehabilitation

Traditional rehabilitation approaches have been widely used for many decades and have had a significant impact on the restoration of functional status and the recovery process of people with PD. The primary strength of traditional rehabilitation approaches is therapist-oriented, physical interaction, manual assessment, and direct supervision. Despite their significant impact on the recovery process, traditional rehabilitation approaches have certain limitations, including the lack of objective assessment, dosage ambiguity, limited accessibility, and a lack of high-intensity, longer-duration training programs. Recently, due to an ageing population and rapid technological advances, there has been a paradigm shift in rehabilitation practices. This shift is important to address the high burden of physical disabilities and to integrate emerging technologies to support the delivery of interventions. The limited number of train staff, along with insufficient healthcare facilities, has placed a significant burden on health systems to provide effective care. Generic healthcare awareness of the population through technologies has increased the demand for uniform, accessible, customised, and cost-effective rehabilitation for all people with PD. These multiple forces paved the way for an innovative, technology-assisted rehabilitation approach (Ali et al., 2024; Mennella et al., 2023b; Willingham et al., 2024).

The current healthcare system, through the traditional rehabilitation model, cannot meet the growing demands because it predominantly relies on therapist-oriented, hospital-based, restricted-access care and a limited workforce. Traditional rehabilitation models are unable to provide intensive, long-term follow-up and a highly repetitive, customised plan to achieve functional recovery. Long-term, sustainable rehabilitation approaches are essential to support the delivery of evidence-based care to people with chronic disabilities and in remote areas. Recent advances in the digitalisation of the healthcare system provide accessible, longer-duration, cost-effective, and customised rehabilitation programs (Mennella et al., 2023b). Emerging technologies play a vital role in addressing the constraints of traditional rehabilitation approaches and in providing vibrant, innovative technology-integrated solutions to achieve meaningful outcomes. The emerging technologies for the rehabilitation of individuals with PDs offer highly intensive and repetitive rehabilitation programs. They promote access to remote areas and address geographic and economic barriers. These technologies also enhance personalised, customised care for all individuals, facilitating long-term engagement in care (Naicker et al., 2019a). In the digital era, rapid technological advancements have been incorporated into rehabilitation to achieve faster recovery. There is an urgent and dire need to provide technologically based rehabilitation services to address the existing challenges and provide accessible, cost-effective, and quality care to people with physical disabilities (Cieza et al., 2020a; Jette, 2021).

1.5 Spectrum of Technologies in Modern Rehabilitation

Technology-based rehabilitation interventions enhance assessment, intervention, monitoring, and long-term functional recovery in patients with physical disabilities. These technologies range from immersive digital environments to intelligent, data-driven systems that address the challenges associated with traditional rehabilitation approaches.

1.5.1 Robotics

Robotics devices are designed to provide intensive, repetitive, and goal-oriented interventions for people with physical disabilities. Robotics is a combination of multiple structures, including mechanical devices, sensors, actuators, and other relevant parts, that interact precisely with the human body. The sensors are highly sensitive for recording kinematic and kinetic feedback from movement and guide control systems to adjust assistance as required. Real-time feedback is one of the most distinctive features of robotics, actively engaging the person in task performance through immediate guidance. The robotic feedback-based training promotes neural plasticity and motor relearning to achieve functional recovery and independence (Liu et al., 2022). Robotic technology is beneficial for facilitating movement by adjusting the dosage of interventions. Robot-assisted therapy is highly effective and recommended in the early stages of recovery, when patients are unable to perform active, independent movements. Robotics is widely used to address gait impairments, postural imbalances, upper-extremity dysfunction, and specific hand impairments, assisting functional rehabilitation. Robotics is commonly used in neuro rehabilitation, but emerging applications are also being adopted in musculoskeletal, geriatric, and paediatric care, where they assist with movement, balance, and gait training (Banyai & Brişan, 2024).

There are two main types of robotics: exoskeletons and end effectors.

Exoskeletons are wearable devices that perform functions through human joints, control movement, and provide support within a specific range of motion. An exoskeleton allows specific exercises at individual joints without affecting other joints of the body. Upper-limb exoskeletons have a mechanical structure, with users connected at different sites, and mimic joint movements. It has a delicate control system to perform smooth execution of movements (Bhujel & Hasan, 2023). Exoskeleton devices are effective for people with physical disabilities who have mobility restrictions and are beneficial for restoring walking abilities. Independence in function and mobility leads to a better quality of life and enhances community participation of people with physical disabilities (Kumar et al., 2023). Exoskeleton devices for upper-limb function assistance, including the ARMin exoskeleton, Hand-SOME, and robotic gloves, are commonly available and effective. Devices for lower-limb function and gait assistance include Lokomat, ReWalk, BLEEX, AUTON-OMYO, PH-EXOS, HUMA, and HAL (Banyai & Brişan, 2024; Hsu et al., 2021). The advantages of exoskeleton devices include precise control of specific joints, flexibility for training, full range of motion, ease of use, and support for intensive training. However, the complex nature, time-consuming design, heavy weight, and power failures pose challenges for the effective implementation of exoskeleton devices. Limited degree of freedom, pain due to malalignment, and high cost are limitations associated with the integration of exoskeleton devices in rehabilitation settings (Banyai & Brişan, 2024; Bhujel & Hasan, 2023).

End-Effector System has been developed for the distal part of the limbs, and joint alignment with the robot and patient is not required for function. It uses different footplates or handles for producing motion of the extremity in space. This system interacts with the human body only far from the limbs, and the movement of one extremity does not affect the other extremities. Common devices include cameras, assembly tools, sensors, and grippers. These devices have a simple structural design, a limited control system, low cost, easy maintenance, less weight, and precise fine-motion control. The challenges include a lack of customised therapy, limited interaction between robots and humans, limited degrees of freedom, limited trajectory, less stability, and limited control of movement. Overall, robotics provides a comprehensive approach with intensive, repetitive practice, a uniform protocol, real-time feedback, and active patient engagement (Demeco et al., 2023). Robotics has the capacity to support high-repetition, high-intensity, high-frequency, and long-duration training for people with physical disabilities. These devices provide quantitative, objective activity data and real-time progress for all individuals. Robotics assists with long-duration training, reducing the physical burden on the therapist, and provides standardised, consistent, and precise training descriptions. Engagement and immediate feedback enhance adherence to training and increase motivation to achieve specific functional recovery in people with physical disabilities (Banyai & Brişan, 2024; Garcia-Gonzalez et al., 2022; Liu et al., 2022).

1.5.2 Virtual Reality and Augmented Reality

Virtual reality (VR) is an interactive computer-generated system that uses sensors to detect human movement and provides virtual environments for interacting with the real world. In 1960, Morton introduced the concept of VR with the creation of the Tele-sphere Mask and Sensorama. Multiple VR-related devices have been developed with control applications. In 2012, Palmer Luckey designed the Oculus prototype, which gained popularity. In 2014, Facebook acquired Oculus for gaming purposes (Hamad & Jia, 2022). The key components of a VR system are hardware, software, human interaction, and virtual presence. The hardware includes headsets, motion sensors, and motion controllers for bridging the connection between humans and VR systems. The software components include virtual activities, avatars, and a virtual environment to provide a suitable space for interaction. Human interaction is a key component in which people use and manipulate objects through visual, sensory, and haptic feedback from the VR system. Another component is the virtual presence and the feeling of humans within the environment created by the VR system (Kouijzer et al., 2023).

VR has been widely used in phobic treatment, military training, and architectural design, but for the last few decades, it has been experimented with in the medical and rehabilitation fields. VR provides real-time sensory, visual, and haptic feedback to monitor and guide the performance of accurate tasks appropriately. VR platform provides an enriching environment to promote neural plasticity, motor relearning, and functional recovery in people with physical disabilities. The gamified nature of VR ensures patient safety when performing complex tasks and exercises that are not possible in a real environment. The virtual presence of patients in a specifically generated environment enhances motivation and emotional engagement toward achieving the desired rehabilitation goals (Demeco et al., 2023; Kouijzer et al., 2023). VR supports experiential and active learning for people with physical disabilities by enabling them to perform activities in virtual environments and providing an experience comparable to real-world ones (Kiani et al., 2023). Three types of VR are commonly used to engage the participants.

Non-Immersive VR uses a two-dimensional design and is displayed on a computer display or a gaming console. Participants interact with a non-immersive virtual environment using various tools displayed on screens. Tools include a mouse, joystick, cyber gloves, cyber grasp, or force sensors that are used for interaction. This system has no immersive features and is suitable for basic simulations, serious games, and cognitive rehabilitation.

Semi-Immersive VR creates three-dimensional images through projections or displays with fixed visual perspectives. Participants experience a deeper sense of connection in a stimulated environment than in a non-immersive one. These systems have large display screens, an automatic virtual environment (CAVE) system, and motion sensors, and are recommended in cognitive and rehabilitation training.

Fully Immersive VR System manages ego-centrically within the surrounding environment through a simulated world. VR devices offer a variety of options for real-time interaction with visual content via head movements. Images can be seen on a head-mounted display, a large-screen projection, or a CAVE. Fully Immersive system comprised of a VR headset, a motion-tracking system, and a motion controller to ensure a real environment and a strong feeling of presence. Immersive VR is widely used in medical education, mental health, and rehabilitation programs (Demeco et al., 2023; Dhar et al., 2023).

Augmented reality (AR) is technology that overlays digital content, including text, images, and graphics, into real-world environments in real-time, allowing the person to interact with both the real and virtual environments simultaneously. Unlike VR, AR does not replace the real world; rather, it promotes human interaction through virtual and physical elements. The three main components of AR are real and virtual objects, real-time human interaction, and precise interaction with physical and digital content.

AR operates by integrating a computer system, sensors, cameras, and a real-time data processing system. AR is widely used in medical, surgical, preoperative guidance, rehabilitation, and patient education in remote areas. AR provides gamification and task-specific exercises that enhance high-repetition, intensive training in people with physical disabilities (Dargan et al., 2023; Phan et al., 2022). Overall, VR provides a safe, risk-free, and protected virtual environment for complex training and activities. The components of gamification and interactions lead to higher patient engagement and long-term adherence in community or home-based rehabilitation for people with physical disabilities. VR provides options for customised games or activities tailored to patients' requirements and their level of physical disability. VR enables high-intensity, repetitive, and standardised activities for longer durations and is useful for remote training and telerehabilitation, reducing therapists' dependence. The VR system also gathers objective-oriented, precise, and accurate performance data for assessment, progress, and monitoring.

1.5.3 Wearable Sensors (WS)

WS were designed as hardware devices in the twentieth century and used for data interaction. Later, users used these small devices as needed to record health-related data. WS has only two parts: the body device and the sensor components, and both are connected. Multiple devices are used to monitor patients' data, including electromyography (EMG), electrocardiogram (ECG), and electroencephalogram (EEG). The rapid development of technology has enabled the integration of emerging technologies with traditional approaches to augment rehabilitation for people with physical disabilities. Wearable sensors (WS) are commonly used in medicine and rehabilitation and assist clinicians in recording progress, monitoring, and predicting recovery across different conditions. In Rehabilitation, different WS are used to track movement, especially gait-related changes, including accelerometers, pressure sensors, and gyroscopes. The Inertial sensor is used to detect and measure acceleration, vibration frequency, rotation angle, tilt, and motion in multiple degrees of freedom. WS devices can convert these movement-related signals to electric signals for circuits. The gyroscope is a motion-detection device that measures angular motion and velocity along different axes, while accelerometer sensors measure velocity changes in a single direction. Pressure sensors are used to detect and measure pressure at a single point on the body and convert these signals into electrical signals. WS are worn on the body at different sites, including the wrist, chest, shoes, or clothes, to collect physiological and movement-related parameters. WS has characteristics that warrant monitoring, assessment, and intervention in health care and rehabilitation. WS monitors posture, physical activity, balance, gait, fall detection, abnormal movements, progression, and home-based rehabilitation for people with physical disabilities (Szabo et al., 2023).

WS provides accurate, objective data and enables early detection and personalised monitoring of activities. WS are beneficial in rehabilitation as they provide home-based training plans, reduce travel strain, and decrease the burden on hospital or clinical settings. WS devices are Non-invasive, user-friendly, and provide real-time feedback to both patients and therapists for monitoring progress and planning (Latif et al., 2024; Lobo et al., 2024; Wei & Wu, 2023a).

1.5.4 Artificial Intelligence (AI)

Artificial intelligence (AI) comprises technologies that simulate human intelligence by making machines perform functions such as decision-making, problem-solving, and pattern recognition. The application of AI-based technologies in health care and rehabilitation is increasingly recognised for its improved assessment, monitoring, diagnosis, and patient management. It can analyse large datasets, detect movement patterns, provide real-time feedback, predict outcomes, enhance patient engagement, increase motivation, and tailor interventions to patients' needs (Alshami et al., 2025).

AI is integrated with technologies such as robotics and exoskeletons, wearable sensors, VR and AR, mobile health applications, and telerehabilitation platforms to enhance patient care. Integration with robots and exoskeletons can help with tasks such as mobility and gait training. When integrated with wearable sensors, it can continuously monitor physiological parameters and movement patterns. AI-enhanced VR/AR systems support exercise training, e.g. balance training and cognitive rehabilitation. Integration with smartphone applications and telerehabilitation platforms enables remote assessment, monitors adherence, and provides feedback to patients and clinicians (Pancholi et al., 2024). Key AI technologies important in rehabilitation include Machine learning (ML), Deep Learning (DL), Computer Vision (CV), and Natural Language Processing (NLP), which allow intelligent systems to process data, learn from experience, and perform complex tasks with minimal human intervention (Liu & Xie, 2025).

Machine learning (ML) is a field of computer science that uses algorithms to analyse data and make predictions or draw conclusions. ML is a subtype of artificial intelligence that enables systems to learn from diverse data, improve performance, and make predictions. ML follows a structured program of data preparation, model training, and evaluation. The raw data is collected from hardware such as wearable sensors, including EMG signals for muscle activity and IMU movement data. Later, the data is normalised to prepare for the algorithm. Time-domain and frequency-domain data have been extracted from the raw data using different calculation methods. However, in deep learning, a convolutional neural network (CNN) does not require manual data extraction. ML has two categories: Supervised learning, in which the data are labelled and categorised, and prediction of outcomes is derived from this data. Unsupervised Learning, in which data is not labelled, and new patterns are identified from this data (Wei & Wu, 2023b).

1.5.5 Brain–Computer Interface (BCI)

BCI is a technology that enables direct communication between the brain and an external device by translating neural activity into control signals, without relying on peripheral nerves or muscles. The BCI system operates within a closed-loop framework and follows the sequence of brain signal acquisition, signal processing & decoding, command translation, and output, with feedback. Brain signals are commonly recorded with EEG, and the output is delivered via robotic exoskeletons, FES, or visual avatars. In rehabilitation, BCIs are used to promote neural plasticity and enhance motor function, sensory integration, and communication in patients with physical disabilities. BCI is shifted from treating impairments at different regions to direct interfacing with the brain to promote motor function. It provides intention-driven movement where there is none and takes control of movement, like a driver. BCI augments active engagement in individuals with no voluntary movement due to physical disabilities and provides a direct neural connection for movement facilitation. BCI Integrates with other technologies, such as robotics, VR, and FES, to enhance neural plasticity and motor relearning in people with physical disabilities. BCI provides customised, personalised training based on individual capabilities and the level of physical disability (Ahmed et al., 2025).

1.5.6 Telerehabilitation (TR)

TR is the delivery of different rehabilitation services in remote/distant areas through information and communication technologies. It provides a platform for health care to educate, prevent, assess, diagnose, treat, and monitor patients in remote areas without being physically present. TR operates through information and communication technologies that connect the therapist and patient for interaction. The technologies include video conferencing, mobile apps, wearable sensors, web-based platforms (Elashmawi et al., 2024), and other digital media systems that are used to provide services. The devices include smartphones, computers, and internet connections for data transformation. TR can be used in both synchronous modes, which support real-time interaction between the patient and therapist, and in asynchronous modes, where recorded video or pictures are used for consultation. Hybrid methods, both live and recorded, can improve efficacy. TR requires a communication, monitoring, and feedback system to provide a suitable environment to manage complex conditions. TR is an effective mode of training for people with physical disabilities for long-term rehabilitation, home-based training, and in remote areas. TR services reduce travel costs, provide real-time feedback, adopt a patient-centred approach, decrease the burden on clinical settings, and are appropriate during a pandemic (Kocyigit et al., 2024; Nicolas et al., 2024).

1.6 Technology Integration in Physical Disability Rehabilitation

The rehabilitation practice has undergone a structural transformation in almost all diagnostic conditions with the integration of technology. The technological systems are fully embedded in rehabilitation protocols for physical disabilities linked with various health conditions. The shared principles of technology-enabled rehabilitation for physical disabilities arising from neurological, musculoskeletal, paediatric, geriatric, or women's health include enhanced motor learning, real-time biofeedback, data-driven personalisation, therapeutic dose intensification, and extending care beyond clinical environments. This section explains how these technologies redefine disability management across major rehabilitation domains.

1.6.1 Neurological Disabilities

The rehabilitation of neurological diseases, disorders, or dysfunctions originating from the central or peripheral nervous systems results in complex impairments, including poor motor control, coordination, sensory integration, and cognition. The traditional principles of rehabilitation in this population rely on repetitive, task-specific training to induce neuroplasticity. Technology integration facilitates this process by increasing intensity, precision, and the availability of feedback. The use of Robotic exoskeletons in neurological patients enhances their motor performance by delivering consistent, high-repetition, reproducible, and task-oriented movement stimuli while reducing therapist burden. Virtual reality platforms, through immersive and non-immersive task simulations, offer an enriched, multisensory environment that promotes motor learning in patients. Further, the cortical engagement and organisation in this population is amplified by Real-time visual and proprioceptive feedback. Similarly, the tele-neurorehabilitation platforms facilitate continued therapy after patient discharge from inpatient services. Sensory-driven feedback systems, including brain–computer interfaces, are also useful tools for disability management, as they directly link neuronal activity to motor output in severe impairments. So, technology serves a pivotal role in disability management across neurological rehabilitation, not merely as an assistive adjunct but as a neuroplasticity amplifier, boosting dose-response relationships and supporting continuous functional tracking.

1.6.2 Musculoskeletal Disabilities

Musculoskeletal disorders or dysfunctions are biomechanical contributors to physical disabilities. The disabling impairments commonly associated with chronic neck pain, back pain, and osteoarthritis are pain, mechanical dysfunction, muscular weakness, and reduced mobility. The highly prevalent disability linked with these disorders can be reduced by the use of rehabilitation technologies in the management protocols and guidelines. The key principles of musculoskeletal disorder training, encompassing chronic pain modulation and precision movement training, can be facilitated by the inclusion of emerging technologies. Musculoskeletal rehabilitation requires specific, tailored exercise prescriptions that are based on objective assessments. For example, wearable biomechanical sensors measure joint angles in real-time, as well as kinetics and kinematics, gait symmetry, and loading patterns. This quantitative data allows clinicians to modify exercise plans with greater specificity. Similarly, artificial intelligence algorithms can help clinicians adjust exercise intensity based on performance metrics using adaptive and progression models. The graded exposure paradigms and virtual reality-based distraction can modulate musculoskeletal pain perception and minimise fear avoidance behaviours linked to chronic pain conditions. In this domain, technology integration for disability rehabilitation goes beyond traditional, generalised strengthening protocols to objective, data-driven, movement-optimised strategies.

1.6.3 Paediatric Disabilities

Paediatric disability rehabilitation in conditions such as cerebral palsy and developmental disorders relies mainly on neurodevelopmental plasticity, which is shaped by motivation and participation. Besides therapeutic intensity, technology plays a distinct role by engaging the child in therapy sessions. The family-centred telerehabilitation models extend therapeutic care into home environments, with fewer access barriers and greater caregiver involvement. Gamified virtual reality environments transform repetitive exercises into interactive, play-based challenges for children, increasing their motivation, adherence, and sustained participation. Delayed ambulation is a major functional limitation in cerebral palsy, and it can be supported by assistive exoskeletons and robotic gait training to facilitate repetitive stepping practice. Smart orthoses and sensor-embedded devices monitor and track movement progress and quality outside clinical sessions.

1.6.4 Geriatric Disabilities

The ageing population presents a growing burden of physical disabilities due to increased risk of falls, frailty, sarcopenia, and balance impairment. Emerging technologies support proactive fall-risk mitigation by reinforcing participation and independence. The immersive virtual reality-based balance training environment exposes them to dynamic balance challenges and perturbation-based exercises to improve postural control while maintaining safety. Early identification and continuous gait monitoring can be achieved using wearable fall detection systems and inertial sensors.

Smart home technologies and telerehabilitation platforms ensure continuity of care for older adults with accessibility barriers to healthcare setups.

1.6.5 Women's Health-Related Disabilities

There are certain physical disabilities associated with health conditions which are either unique, more prevalent, present more seriously, carry different risk factors, or are treated differently in women. These include postmenopausal osteoporosis, breast cancer-related lymphedema, and post-gynaecological surgeries linked to pelvic floor dysfunction. Technologies such as AI-enabled wearable sensors support early detection of lymphedema progression and adherence to personalised exercise protocol. Smart compression textiles and adaptive garments represent emerging therapeutic innovations that enhance function with comfort and psychosocial acceptability. Many technological interventions are effective for pelvic floor rehabilitation, including biofeedback devices and digital perineometers, which enable remote monitoring and guided exercise progression. This population experiences limited access to health care due to caregiving responsibilities at home and social restrictions. So, telehealth platforms can serve to address the unmet healthcare needs of women.

1.7 Conclusion

The world is bearing an exponential rise in physical disabilities driven by an ageing population, enhanced survival rate, increased prevalence of non-communicable chronic diseases and disorders. Our traditional rehabilitation models are insufficient to meet the complex and growing demand for disability care. The integration of emerging technologies into disability rehabilitation is a fundamental paradigm shift that provides us with a modern and evolving framework of care for PWD. The advent of this modern rehabilitation model opens the way for innovative, personalised, precise, sustainable, and accessible care for people with disabilities. The transition is characterised by a shift from clinic-based, episodic patient-clinician interactions to continuous, data-informed monitoring and decision-making, and to extended care beyond rehabilitation settings. The spectrum of technologies discussed, such as robotics, virtual and augmented reality, wearable sensors, artificial intelligence, brain–computer interface, and telerehabilitation platforms, is introducing greater objectivity, precision, and accessibility to disability-specific therapeutic interventions. Such integrations across neurological, musculoskeletal, paediatric, geriatric, and women's health conditions enhance outcomes across all levels of disability, including impairments, activity performance, participation, and overall quality of life. The inclusive principles of technology-based rehabilitation exacerbate existing geographic and socioeconomic disparities and redefine how physical disabilities are managed across diverse health systems worldwide.

References

Ahmed, B., Khan, S., Lim, H., & Ku, J. (2025). Challenges and opportunities of gamified BCI and BMI on disabled people learning: A systematic review. *Electronics, 14*(3), 491.

Ali, W. W., Kassim, A. A., & Rashid, S. M. (2024). Assistive technology for persons with physical disabilities: A literature review. *International Journal of Academic Research in Business and Social Sciences, 14*(2), 544–560.

Alshami, A., Nashwan, A., AlDardour, A., & Qusini, A. (2025). Artificial Intelligence in rehabilitation: A narrative review on advancing patient care. *Rehabilitacion, 59*(2), 100911. https://doi.org/10.1016/j.rh.2025.100911

Banyai, A. D., & Brişan, C. (2024). Robotics in physical rehabilitation: Systematic review. *Health Care, 12*(17), 1720.

Bhujel, S., & Hasan, S. (2023). A comparative study of end-effector and exoskeleton type rehabilitation robots in human upper extremity rehabilitation. *Human-Intelligent Systems Integration, 5*(1), 11–42.

Cieza, A., Causey, K., Kamenov, K., Hanson, S. W., Chatterji, S., & Vos, T. (2020a). Global estimates of the need for rehabilitation based on the Global Burden of Disease study 2019: A systematic analysis for the Global Burden of Disease Study 2019. *The Lancet, 396*(10267), 2006–2017.

Cieza, A., Causey, K., Kamenov, K., Hanson, S. W., Chatterji, S., & Vos, T. (2020b). Global estimates of the need for rehabilitation based on the Global Burden of Disease study 2019: A systematic analysis for the Global Burden of Disease Study 2019. *The Lancet, 396*(10267), 2006–2017.

Dargan, S., Bansal, S., Kumar, M., Mittal, A., & Kumar, K. (2023). Augmented reality: A comprehensive review: S. Dargan et al. *Archives of Computational Methods in Engineering, 30*(2), 1057–1080.

Demeco, A., Zola, L., Frizziero, A., Martini, C., Palumbo, A., Foresti, R., Buccino, G., & Costantino, C. (2023). Immersive virtual reality in post-stroke rehabilitation: A systematic review. *Sensors, 23*(3), 1712.

Dhar, E., Upadhyay, U., Huang, Y., Uddin, M., Manias, G., Kyriazis, D., Wajid, U., AlShawaf, H., & Syed Abdul, S. (2023). A scoping review to assess the effects of virtual reality in medical education and clinical care. *Digital Health, 9* https://doi.org/10.1177/20552076231158022

Elashmawi, W. H., Ayman, A., Antoun, M., Mohamed, H., Mohamed, S. E., Amr, H., Talaat, Y., & Ali, A. (2024). A comprehensive review on brain–computer interface (BCI)-based machine and deep learning algorithms for stroke rehabilitation. *Applied Sciences, 14*(14), 6347.

Garcia-Gonzalez, A., Fuentes-Aguilar, R. Q., Salgado, I., & Chairez, I. (2022). A review on the application of autonomous and intelligent robotic devices in medical rehabilitation. *Journal of the Brazilian Society of Mechanical Sciences and Engineering, 44*(9), 393.

Gimigliano, F., & Negrini, S. (2017). The World Health Organization "rehabilitation 2030: A call for action". *European Journal of Physical and Rehabilitation Medicine, 53*(2), 155–168.

Hamad, A., & Jia, B. (2022). How virtual reality technology has changed our lives: An overview of the current and potential applications and limitations. *International Journal of Environmental Research and Public Health, 19*(18), 11278.

Hsu, S.-H., Changcheng, C., Lee, H.-J., & Chen, C.-T. (2021). Design and implementation of a robotic hip exoskeleton for gait rehabilitation. *Actuators, 10*(9), 212.

Jette, A. M. (2021). Global prevalence of disability and need for rehabilitation. *Physical Therapy, 101*(2), pzab004.

Kiani, S., Rezaei, I., Abasi, S., Zakerabasali, S., & Yazdani, A. (2023). Technical aspects of virtual augmented reality-based rehabilitation systems for musculoskeletal disorders of the lower limbs: A systematic review. *BMC Musculoskeletal Disorders, 24*(1), 4.

Kocyigit, B. F., Assylbek, M. I., & Yessirkepov, M. (2024). Telerehabilitation: Lessons from the COVID-19 pandemic and future perspectives. *Rheumatology International, 44*(4), 577–582.

Kołłątaj, B., Kołłątaj, W. P., Cipora, E., Piecewicz-Szczęsna, H., & Karwat, I. D. (2025). Determination of the medical and social levels of disability as a basis for successful comprehensive rehabilitation. *Annals of Agricultural and Environmental Medicine, 32*(2), 198–205.

Kouijzer, M. M., Kip, H., Bouman, Y. H., & Kelders, S. M. (2023). Implementation of virtual reality in healthcare: A scoping review on the implementation process of virtual reality in various healthcare settings. *Implementation Science Communications, 4*(1), 67.

Kumar, J., Patel, T., Sugandh, F., Dev, J., Kumar, U., Adeeb, M., Kachhadia, M. P., Puri, P., Prachi, F., & Zaman, M. U. (2023). Innovative approaches and therapies to enhance neuroplasticity and promote recovery in patients with neurological disorders: A narrative review. *Cureus, 15*(7).

Kumurenzi, A., Richardson, J., Thabane, L., Kagwiza, J., Urimubenshi, G., Hamilton, L., Bosch, J., & Jesus, T. (2023). Effectiveness of interventions by non-professional community-level workers or family caregivers to improve outcomes for physical impairments or disabilities in low resource settings: Systematic review of task-sharing strategies. *Human Resources for Health, 21*(1), 48.

Latif, A., Al Janabi, H. F., Joshi, M., Fusari, G., Shepherd, L., Darzi, A., & Leff, D. R. (2024). Use of commercially available wearable devices for physical rehabilitation in healthcare: A systematic review. *BMJ Open, 14*(11), e084086.

Liu, C., Lu, J., Yang, H., & Guo, K. (2022). Current state of robotics in hand rehabilitation after stroke: A systematic review. *Applied Sciences, 12*(9), 4540.

Liu, H., & Xie, Q. (2025). Applications of artificial intelligence in rehabilitation: Technological innovation and transformation of clinical practice. *SLAS Technology*, 100360. https://doi.org/10.1016/j.slast.2025.100360

Lobo, P., Morais, P., Murray, P., & Vilaça, J. L. (2024). Trends and innovations in wearable technology for motor rehabilitation, prediction, and monitoring: A comprehensive review. *Sensors, 24*(24), 7973. https://doi.org/10.3390/s24247973

Mennella, C., Maniscalco, U., De Pietro, G., & Esposito, M. (2023a). The role of artificial intelligence in future rehabilitation services: A systematic literature review. *IEEE Access, 11*, 11024–11043.

Mennella, C., Maniscalco, U., De Pietro, G., & Esposito, M. (2023b). The role of artificial intelligence in future rehabilitation services: A systematic literature review. *IEEE Access, 11*, 11024–11043.

Naicker, A. S., Htwe, O., Tannor, A. Y., De Groote, W., Yuliawiratman, B. S., & Naicker, M. S. (2019a). Facilitators and barriers to the rehabilitation workforce capacity building in low-to middle-income countries. *Phys Med Rehabil Clin N Am, 30*(4), 867–877.

Naicker, A. S., Htwe, O., Tannor, A. Y., De Groote, W., Yuliawiratman, B. S., & Naicker, M. S. (2019b). Facilitators and barriers to the rehabilitation workforce capacity building in low-to middle-income countries. *Physical Medicine and Rehabilitation Clinics of North America, 30*(4), 867–877.

Neill, R., Shawar, Y. R., Ashraf, L., Das, P., Champagne, S. N., Kautsar, H., Zia, N., Michlig, G. J., & Bachani, A. M. (2023). Prioritizing rehabilitation in low-and middle-income country national health systems: A qualitative thematic synthesis and development of a policy framework. *International Journal for Equity in Health, 22*(1), 91.

Nicolas, B., Leblong, E., Fraudet, B., Gallien, P., & Piette, P. (2024). Telerehabilitation solutions in patient pathways: An overview of systematic reviews. *Digital Health, 10*. https://doi.org/10.1177/20552076241294110

Pancholi, S., Wachs, J. P., & Duerstock, B. S. (2024). Use of artificial intelligence techniques to assist individuals with physical disabilities. *Annual Review of Biomedical Engineering, 26*(1), 1–24.

Phan, H. L., Le, T. H., Lim, J. M., Hwang, C. H., & Koo, K. (2022). Effectiveness of augmented reality in stroke rehabilitation: A meta-analysis. *Applied Sciences, 12*(4), 1848. https://doi.org/10.3390/app12041848

Szabo, D. A., Neagu, N., Teodorescu, S., Apostu, M., Predescu, C., Pârvu, C., & Veres, C. (2023). The role and importance of using sensor-based devices in medical rehabilitation: A literature review on the new therapeutic approaches. *Sensors, 23*(21), 8950.

Wei, S., & Wu, Z. (2023a). The application of wearable sensors and machine learning algorithms in rehabilitation training: A systematic review. *Sensors, 23*(18), 7667. https://doi.org/10.3390/s23187667

Wei, S., & Wu, Z. (2023b). The application of wearable sensors and machine learning algorithms in rehabilitation training: A systematic review. *Sensors, 23*(18), 7667. https://doi.org/10.3390/s23187667

Willingham, T. B., Stowell, J., Collier, G., & Backus, D. (2024). Leveraging emerging technologies to expand accessibility and improve precision in rehabilitation and exercise for people with disabilities. *International Journal of Environmental Research and Public Health, 21*(1), 79.

Chapter 2
Emerging Technologies in Neurological Rehabilitation

Arshad Nawaz Malik⬡

Abstract Neurological disorders are gradually rising globally and are the leading cause of long-term disability, creating a serious healthcare challenge. Impairment and disability associated with these conditions create a significant economic, social, and healthcare burden on the community. Neurological disorders such as Stroke, Parkinson's disease, Spinal Cord Injury, Traumatic Brain Injury, Multiple Sclerosis, and Amyotrophic Lateral Sclerosis cause motor and cognitive impairment, with limited functional outcomes. These impairments lead to restricted activity and participation, a decline in quality of life, and increased long-term dependence. Conventional rehabilitation is crucial for restoring functional recovery, but it is limited by insufficient, inaccessible, and engaging training. Emerging technologies are innovative approaches that provide an accessible, intensive, and feedback-oriented plan to enhance effectiveness. This chapter explains the recovery mechanism of neurological disorders, highlighting the key features of motor relearning and neural plasticity. This chapter provides an outline of major emerging technologies, including robotics, virtual and augmented reality, brain–computer interface, wearable sensors, non-invasive brain stimulation, and telerehabilitation platforms. This chapter also evaluates the disease-specific applications of emerging technologies on interventions to enhance functional and motor recovery through patient engagement. Finally, the challenges and future directions of emerging technology have been discussed to develop accessible, personalised, and evidence-based rehabilitation.

Keywords Neurological Disorders · Neurorehabilitation · Robotics · Stroke · Telerehabilitation · Virtual Reality

A. N. Malik (✉)
Faculty of Rehabilitation and Allied Health Sciences, Riphah International University, Islamabad, Pakistan
e-mail: arshad.nawaz@riphah.edu.pk

© The Author(s), under exclusive license to Springer Nature Singapore Pte Ltd. 2026
A. N. Malik et al., *Emerging Technologies in the Rehabilitation of Physical Disabilities*, SpringerBriefs in Modern Perspectives on Disability Research, https://doi.org/10.1007/978-981-92-1343-6_2

2.1 Global Burden and Classification of Neurological Disorders

Neurological disorders (NDs) are the leading cause of disability and are rapidly developing into a significant health challenge worldwide. NDs are a diverse group of diseases that affect the brain, spinal cord, and peripheral nerves. NDs have different pathophysiology mechanisms of diseases, such as vascular impairments, including ischemia and haemorrhage, neurodegenerative, autoimmune demyelination, traumatic injuries, infections, and genetic metabolic abnormalities. The disruption in blood flow, damage to synaptic connections, abnormal protein deposition, and altered nerve conduction collectively lead to neuronal and functional loss. The clinical impairments associated with NDs include motor impairments (spasticity, paralysis, tremors), sensory loss, cognitive impairments, communication loss, and autonomic dysfunctions. In most of the NDs, long-term physical disability reduces the functional level, independence, activity status, and social participation of affected people. The nature of individuals' dependence on family and society for a longer duration poses a significant economic and social burden on the community (Feigin et al., 2020a).

The prevalence of NDs has increased significantly globally over the last two decades. The global rise in population, ageing, and lifestyle changes is contributing to the high prevalence of these conditions. The central driver of this rapid increase is population ageing, which is directly linked to NDs. Sedentary lifestyle, urbanisation, obesity, diabetes, hypertension, smoking, and environmental exposure are contributing factors and play a critical role in both high- and low-income countries. According to Global Burden of Disease (GBD) reports, 18 major NDs collectively account for approximately 10.2% of global disease burden, and almost 16.8% of all deaths worldwide. NDs are responsible for nearly 349.42 million Disability-Adjusted Life Years (DALYs). NDs are the leading cause of disability globally and pose a significant challenge to healthcare systems in low-resource countries. The limited access to early diagnosis, acute care, rehabilitation services, and community-based services intensifies long-term disability (Lima et al., 2022).

Neurological Disorders are broadly divided into three categories based on aetiology and disease characteristics.

Non-Communicable Diseases include Stroke, Parkinson's disease, Alzheimer's disease, Multiple Sclerosis, and Motor Neuron Disease.

Communicable Diseases include Encephalopathy, Meningitis, and Tetanus.

Neurological Injuries include Traumatic Brain Injury (TBI) and Spinal Cord Injury (SCI) (Ding et al., 2022a).

2.2 Impact of Neurological Impairments and Functional Limitations

Neurological disorders have diverse and complicated impairments that severely disrupt the body's structure & function, limit activities of daily life, and restrict participation within the community. The central and peripheral nervous systems control and coordinate everything from simple reflexive movements to complex, task-specific, and high-performance motor activities. Damage to the nervous system disrupts the integration of sensory stimuli, motor outcomes, and higher cognitive processing, which is crucial for planning and executing controlled movements. Consequently, this interface impairs cognitive, sensory, motor, perceptual, psychological, and executive brain functions (Feigin et al., 2020a). The major impairments include paralysis or paresis, spasticity, tremors, loss of coordination, bradykinesia, loss of selective motor control, involuntary movement, and abnormal movement patterns. Sensory impairments, such as loss of kinesthesia, proprioception, touch, and discriminative sensations, as well as pain, reduce motor control and functional independence. Cognitive impairments include attention deficits, memory loss, apraxia, poor planning, and impaired decision-making, leading to a decline in problem-solving ability and judgement. Psychological disturbances include emotional issues, anxiety, and depression, which further hinder communication, active participation, and social interaction. Collectively, these impairments lead to poor functional performance, dependence on family, limited activities of daily living, and restricted participation in instrumental activities of daily living, such as employment and community engagement (Feigin et al., 2020a).

The long-term physical disability associated with neurological disorders creates a considerable burden on hospitals, rehabilitation centres, and community care centres due to the longer duration of intervention and care. Physical disability reduces walking ability, mobility, and self-care, and leads to a lack of appropriate communication, which leads to social isolation, dependence on family, and unemployment. In lower-middle-income countries (LMICs), caregivers and family members without formal training experience depression, physical exertion, emotional disturbances, and financial constraint. The costs of long-term care, frequent travel, medications, and unemployment place serious stress on families. Undue pressure on caregivers is also a challenge and has reduced a family's financial sustainability. Unfortunately, there is no appropriate system of social support for such families during this episode of disability. The insufficient healthcare system is also a serious threat to the high prevalence of physical disability and social burden on the community. There are only 24% of countries having a specific policy for neurological disorders, and only 12% have allocation of budget for taking care of these long-term disabilities. There is a lack of planning to invest in preventive measures, to increase the healthcare workforce, and to establish a community rehabilitation centre to manage these conditions (Wang et al., 2023). In high-income countries, there is a structured system to address disease and disability through diverse plans of action within healthcare and social support. But recently, the lack of professional staff has emerged as a challenge due to population ageing and high prevalence of neurological disorders. Globally, the rising burden of NDs and the lack of proper care services emphasise the urgent need for accessible, cost-effective, and innovative solutions for people with NDs. Socioeconomic impact is a consequence of long-term disability, especially in LMICs (Feigin et al., 2020b).

2.3 Mechanism of Motor Recovery in Neurological Disorders

Neural plasticity, or the brain's plasticity, refers to the brain's ability to reorganise physiological, structural, and functional aspects throughout life. It provides new synaptic formation, collateral pathways, and unmasking of previously unmasked neural circuits. There are two types of plasticity: synaptic and structural. Synaptic plasticity, through long-term potentiation and long-term depression, supports learning. This capacity of the brain enhances new learning, improves memory, and aids recovery from various types of brain injury. In neurological conditions, there is damage to the brain structure and a compromise of function. Structural plasticity provides alternative connections to recover motor & cognitive, as well as higher mental functions. Neural plasticity provides a framework for functional recovery through axonal and dendritic sprouting, the formation of new synaptic connections, and the establishment of collateral pathways (Kumar et al., 2023b). The different therapeutic interventions follow key parameters, including task-oriented training, repetition, concurrent feedback, and sensorimotor stimulation, to provide an enriching environment for neural plasticity, thereby leading to robust functional outcomes. Meaningful change after a brain injury can be achieved through neural plasticity with focused, structured training programs. The capacity of the brain's features to adapt and rearrange the structure to provide alternative pathways for recovery. The evidence-based approaches focused on task-oriented activities, structured feedback, intensive repetition, and active patient engagement. It is important to consider the basic principles of neural plasticity in each intervention strategy. The core principle of experience-dependent neural plasticity includes repetition, intensity, specificity, salience, feedback, timing, and transference. The application of these principles is crucial in both conventional and technology-assisted rehabilitation to maximise functional recovery, independence, and participation.

Motor learning is the process of acquiring new knowledge or skills through experience and practice, which leads to structural changes in the brain. Functional recovery in neurological disorders is closely linked to the motor learning process, as it promotes activity and participation performance. Key parameters of the motor learning approach for recovery include task-specific activities, task repetition, training intensity, and feedback. Task-specific concept mainly reflects the gold standard in learning, which is "practice what you have lost or what you would like to gain". The brain understands the movement pattern rather than recognising the muscles involved in task performance. Specific functional movements during task-oriented activities stimulate the neural circuits required for that task. The task-specific activities must be contextual, like real-life tasks, to achieve the targeted goals of rehabilitation. Repetitions are key to neural plasticity and motor learning, leading to effective structural changes that form a new synaptic network. It is important to understand that every skill or activity requires thousands, and even millions, of repetitions to achieve meaningful and excellent task execution. The term intensity has different operational definitions in literature, but it commonly refers to the frequency, duration, and progression of tasks in a rehabilitation plan. It is well established that a highly intensive training regime results in greater recovery than an optimal intensive plan. Training for a longer duration promotes neural plasticity and cortical reorganisation, facilitating early recovery of motor function. Feedback of movements and results is an essential component that provides error correction and enhances motor learning. There are two types of feedback: intrinsic, generated within the body, and extrinsic, generated by the external environment. Concurrent feedback plays a significant role in functional recovery by enhancing motor learning and performance (Kleim & Jones, 2008).

2.4   Limitations of Conventional Neurorehabilitation Approaches

The conventional neurorehabilitation approaches include Neurodevelopmental approaches (Bobath), motor relearning programs, task-oriented training, modified Constraint-Induced Movement Therapy (mCIMT), mirror therapy, mental imagery, proprioceptive neuromuscular facilitation (PNF) techniques, sensory integration, cognitive rehabilitation, and rhythmic auditory stimulation. These neurorehabilitation strategies provide a basic structural interventional plan to achieve optimal motor and cognitive functions. However, these approaches face various limitations that restrict early and accelerated functional recovery. The major limitation is the lack of high repetitions of certain tasks to achieve a meaningful neuroplastic change. Fewer staff and therapists, shorter session duration, and patient fatigue also limit the intensity of the training. Therapist-oriented training sessions require physical exertion and full-time supervision, which affects the workload and the lack of standardised training for all patients.

Active engagement and motivation are crucial aspects for strong adherence to training, but cognitive impairment, mood disturbances, physical fatigue, and monotonous sessions reduce active participation. The lack of appropriate objective documentation, monitoring, and tracking limits real-time feedback to assess performance and provide guidance for improvement. In low-income countries, there is a serious challenge of accessibility, a lack of trained rehabilitation professionals, and fewer rehabilitation and community centres to manage these patients. Additionally, the high cost of long-term care, travel costs, and the burden on caregivers lead to poor recovery. Overall, the lack of repetition, feedback, intensive training, an individualised plan, accessible motivation, and therapist-assisted training contributes to poor functional recovery. It is crucial to address these challenges using emerging rehabilitation technologies to develop the most suitable strategies for neurological disorders (Kwakkel, 2006; Reddy, 2025). Currently, there is an urgent need to develop a comprehensive management plan for physical disabilities associated with neurological disorders. Updated and comprehensive rehabilitation is decisive to address the most prevalent burden of disability and promote fast, effective, and efficient functional recovery. Accessible, early, and appropriate rehabilitation supports the person with physical disabilities to become a functionally independent, productive citizen and perform an occupational role with dignity. Additionally, active participation in the community enhances the quality of life and reduces the short and long-term social and economic burden on society. The integration of emerging technologies with traditional rehabilitation strategies offers cutting-edge, cost-effective, accessible, and comprehensive plans. This integrated strategic approach provides a comprehensive model for implementing innovative technologies to address the limitations of traditional methods.

2.5 Rationale for Emerging Technologies in Neurorehabilitation

Emerging technologies have brought about a paradigm shift in the medical and rehabilitation fields by improving instrument capabilities, enhancing accessibility and interactivity, and pushing beyond boundaries. Technology-assisted rehabilitation offers an opportunity to improve neural plasticity and enhance functional status in patients with neurological diseases. There are a few key components that support the role of emerging technologies in effective rehabilitation, including repetition, an enriched environment, cost-effectiveness, telerehabilitation, and AI integration. These technologies offer solutions to the limitations of conventional rehabilitation, including highly intensive, task-oriented engagement strategies and opportunities for individualised training plans. The advantages of technology-assisted rehabilitation include the smooth delivery of large numbers of repetitions, the integration of function-oriented tasks, and real-time feedback to promote neural plasticity. Therapist-dependent manual training is also a major challenge addressed by these technologies. VR/AR provides a specific, gamified platform for engaging patients in task-specific, repetitive, and progressive training (Proietti et al., 2022).

Enriched Environment is a key factor in neuroplastic changes and cortical reorganisation through diverse experiences and exposures. Emerging technologies offer a distinct environment, including visual, auditory, and haptic feedback, as well as sensory and motor integration. The environment includes non-immersive, semi-immersive, and immersive modes, depending on the training requirements, to support active engagement. The features of the environment include safety, control, monitoring, and resemble real-time activities (Maier et al., 2019).

Cost-effectiveness is one of the long-term advantages of emerging technologies. However, during the initial phase of incorporating emerging technologies, they are costly due to high costs, the need for essential accessories, and space allocation. But in the long term, this is a cost-effective approach that reduces therapists' workload, promotes home-based training, and reduces travel costs. The reduction in patients' hospital or clinic visits also reduces the burden on the healthcare system (Barrett et al., 2024).

Telerehabilitation is a significant advantage of emerging technologies, as it enables training, guidance, monitoring, and feedback for patients in remote areas. TR resolves one of the main constraints on training accessibility for everyone, regardless of distance or financial limitations.

AI integration is also a useful tool for offering individualised training plans for patients based on their needs and progression, with progress tracked against their capabilities. As every patient is different and requires a specific, individualised training protocol, AI integration addresses this gap by generating a person-specific plan.

Overall, emerging technologies offer cutting-edge tools for comprehensive, accessible training programs. Therefore, the incorporation of emerging technologies in the neurological rehabilitation process is duly justifiable. Technology-assisted intervention includes robotics, wearable sensors, virtual reality, augmented reality, and brain–computer interfaces, providing leverage for intensity modifications, repetitions, self-management support, individualised interventions, and enhanced motivation and active participation (Alt Murphy et al., 2024a).

2.6 Emerging Technologies in Neurorehabilitation

Emerging technologies have been used to foster neural plasticity, thereby promoting recovery and functional outcomes in neurological diseases. These technologies play a significant role in comprehensive neurorehabilitation, helping patients with physical disabilities achieve rapid recovery and independence.

2.6.1 Robotic-Assisted Technology

Robot-assisted training (RAT) is the "application of electronic, computerised control systems to mechanical devices designed to actively and interactively assist human function during rehabilitation of upper or lower limbs, either through exoskeleton or end effector types". Robotic-assisted therapy enhances neural plasticity and functional recovery through intensive and task-specific training. Robotics offer a diverse platform for the repetition of movement at higher intensity, achieving the optimal functional target in a limited duration. Robotics provide operational capacity with an unlimited margin, without physical fatigue, and maintains the quality of movement in a uniform, standardised manner. Immediate, real-time feedback, objective assessment, and precise monitoring facilitate active patient engagement in intensive therapy sessions. Robotics also incorporates gamified, interactive virtual environments to encourage patients to adhere to long-term rehabilitation programs (Bhujel & Hasan, 2023).

Robotics provide auditory, visual, and haptic feedback. Haptic feedback provides a sense of touch in some virtual experiences through actuators and delivers touch sensations through vibrations, emotions, and forces into the real world. This sensory integration within the robotics system enhances neural plasticity. These devices have actuators, sensors, and control systems intended to assist movement in patients with mobility dysfunction. In rehabilitation, these devices are used to enhance motor function and facilitate overall physical exercise (Banyai & Brişan, 2024a). There are two main types of robotics, including exoskeletons and end effectors. Overall, robotics provides a comprehensive approach with intensive, repetitive practice, a uniform protocol, real-time feedback, and active patient engagement. However, high costs, complex structures, maintenance constraints, and limited access in low-income countries are challenges to the implementation of neurorehabilitation.

2.6.2   Virtual Reality (VR) and Augmented Reality (AR)

Virtual Reality (VR) technologies create virtual environments and serve as an alternative training platform. Virtual reality can be defined as "a medium composed of interactive computer simulations that sense the participant's position and actions and replace or augment the feedback to one or more senses, giving the feeling of being mentally immersed or present in the simulation (a virtual world)". Three types of VR are commonly used to engage the participants: non-immersive, semi-immersive, and full-immersive. AR is an emerging technology that creates real-time interactive experiences using virtual and real objects. AR provides real-time, accurate feedback to participants, improving performance and enhancing learning. Augmented reality also assists the practice through real-world interaction. It provides a controlled, safe environment for participants to interact with real objects in real environments. This system can adapt activities to patients' needs and preferences. These are effective in improving the functional outcome, motor performance, cognitive function, postural control, and overall mobility in patients with neurological diseases (Leong et al., 2022a).

2.6.3 Brain–Computer Interface (BCI)

A Brain–Computer Interface (BCI) provides a closed-loop, dynamic monitoring system. This system enables component interaction and provides sensory feedback, establishing a direct connection between the brain and computers. BCI activates the ipsilateral motor neuron, leading to desynchronization in the same hemisphere and thus stimulating the same primary and secondary cortex (J. Liu et al., 2025a). There is also a change in the fractional anisotropy of the corticospinal tract via BCI, suggesting improved motor function. BCI is an artificial intelligence-based technology that has recently seen increased use in medicine and rehabilitation. It enhances neural plasticity by directly impacting the brain, promoting new synaptic connections and the reorganisation of function. BCI uses the brain's collateral pathways for communication and bypasses the brain's connections to peripheral muscles and nerves. It facilitates normal motor control, promoting functional recovery at the brain level after stroke (Y. Peng et al., 2022).

2.6.4 Wearable Sensor Devices

Wearable sensors (WS) are small, compact, lightweight, and non-invasive electronic devices designed for specific body parts, such as wrists, chests, and joints, to capture movement data. The most common WS in neurorehabilitation include accelerometers and the inertial measurement units (IMUs). Accelerometers measure motion in one to three linear planes, while IMUs record motion in three-dimensional linear acceleration and angular velocity data from accelerometers and gyroscopes, respectively. The dual combination of linear and angular motion provides a comprehensive picture of upper-limb motion, encompassing multiple degrees of freedom. Accelerometer devices are simple and better options in uncontrolled, remote environments with magnetic interference. IMUs are more sensitive and data-rich devices used for high-performance assessment and measurement (Kim et al., 2022a). WS are the best devices for detecting, tracking, and monitoring various components of neurorehabilitation. The WS helps clinicians make evidence-based decisions through objective-oriented feedback. These devices support the design of an individualised training plan with variations in motor patterns and movements. The precise monitoring system provides a better training approach with progression and optimisation leading to an outcome-oriented program (Kumar et al., 2023a).

2.6.5 Telerehabilitation

Telerehabilitation is a key emerging technological platform to benefit the large-scale population of NDs. It provides an environment in which therapists can connect with patients in remote areas via communication technologies for consultation. TR helps with prevention, assessment, diagnosis, intervention, precautions, monitoring, awareness, and remote tracking of patients' progression. Various telecommunication devices are used to develop suitable interaction between patients and therapists in NDs. TR is an extension of the clinical setting training protocol rather than a complete replacement, to enhance accessibility, reduce travel costs, and be suitable for long-term care. TR works in synchronous, asynchronous, and hybrid modes to facilitate, as resources permit. The role of TR in NDs is well established: it promotes neural plasticity, sensorimotor function, gait and balance outcomes, and overall enhances quality of life. TR represents an appropriate approach for highly intensive training protocols that do not require hands-on intervention. TR is beneficial for chronic conditions, home-based training, remote areas, and patient engagement in mild-to-moderate neurological impairments (Del Pino et al., 2022).

2.7 Disease-Specific Applications of Emerging Technologies

2.7.1 Stroke or Cerebro-Vascular Accident (CVA)

Cerebrovascular Accident (CVA) or Stroke is the leading cause of disability and mortality worldwide. The stroke incidence is 12.2 million, the mortality rate is 6.55 million, and 143.2 million DALYs are the highest among all neurological disorders. Stroke has diverse forms of impairment, including limited upper-limb function, inability to perform daily activities of life, impaired postural control, high risk of falls, balance dysfunction, slow waking speed, and deviated gait pattern (Ding et al., 2022b). These impairments lead to a decline in activity level, participation restrictions, and overall decline in quality of life. Two main strategies for stroke rehab include active training to regain physical function and passive training through orthosis to provide compensation for the loss of function (Calafiore et al., 2022).

The following emerging technologies are commonly used to support early recovery and enhance neural plasticity after stroke.

Robotic-assisted technology is well-suited for stroke rehabilitation, enabling intensive feedback-oriented training. Robot-assisted gait training has a significant impact on stroke rehabilitation through intensive, high-dose, and increased repetition. This intensive and focused training improves postural control, balance, and walking patterns by creating new synaptic connections and reorganising damaged brain areas (Calafiore et al., 2022). Post-stroke walking recovery takes the first 11 weeks, as this is called the "critical rehabilitation period", where there is a high chance of neural plasticity. However, these neuroplastic changes also emerge in chronic stroke after 6 months of stroke. Exoskeleton robotic-assisted training (ERAT) is effective in the subacute phase for gait performance. The most common robotic device for gait training is the Lokomat R (Hocoma AG, Volklstetten, Switzerland). This actuated exoskeleton promotes a symmetrical bilateral gait pattern, equal load distribution, proprioceptive stimulation, and normal gait-cycle phases. This stimulating pattern, with proper load distribution along with kinetic and kinematic integration, promotes neural plasticity and improves brain motor schemas (J. Yang et al., 2023).

Locomat R is a robotic bilateral orthosis that provides automatic locomotion and is equipped with a body-weight support system and a treadmill. It is also associated with an augmented reality system to enhance engagement and increase activity levels. Lokomat R is used to increase training dosage, intensity, and the number of repetitions, as well as task-oriented exercises, to promote recovery. It combines top-down and bottom-up approaches to enhance neural plasticity and functional recovery (Baronchelli et al., 2021). The wearable cyborg Hybrid Assistive Limb (HAL) is the first device to use BioElectrical Signals (BES) from the skin surface to generate muscle force and promote function. The BES is a sensor system that detects the activity of the extensor and flexor muscles around the knee and hip joints. HAL also detects reaction forces, measures joint angles, and measures toe forces. The intervention is based on hybrid control systems that are voluntary and autonomous and uses interactive feedback to promote neural network reorganisation. The interactive biofeedback generated by the HAL system produces a strong learning effect; therefore, this is an innovative approach to promoting independent walking (Watanabe et al., 2021).

The Soft Robotic Gloves (SRGs) are effective for hand function in both the short and long term. Fine-motor skills have shown better results with SRGs, with more than 30 minutes of sessions, in the chronic stage after six months of stroke. Although there is a need for uniform prescription of SRGs in stroke rehabilitation of the upper extremity (Ko et al., 2023). Robot-assisted therapy (RAT) facilitates upper-limb function through repetitive, task-specific training. It benefits all stakeholders, including patients, therapists, healthcare providers, and insurers, by enabling a speedy recovery of function, encouraging independence, and reducing hospital burden. RAT improves dexterity, manages pain, reduces spasticity, and enhances quality of life after stroke. RAT has high potential for early recovery and better outcomes as compared to conventional therapy (Fiore et al., 2023). The protocol for robot-assisted gait training includes 20–60 minutes per session, 3–5 days per week, and a total duration of 4–6 weeks. The most common range is 15–20 sessions and is more effective in the first three months of stroke, when patients are unable to ambulate. Robot-assisted gait training should be incorporated into conventional training to achieve the desired recovery goals. Upper-limb robot-assisted training should be 30–45 minutes, up to 60 minutes, depending on patients' tolerance. The protocol should be provided for 3–5 days per week, and, for intensive training, 5 days is recommended. A total of 4 weeks of duration is recommended for better outcomes, in addition to the conventional hand training protocol (Banyai & Brişan, 2024b).

Virtual Reality (VR) and Augmented Reality (AR) have gained attention in stroke rehabilitation and offer a platform for intensive, repetitive, motivating, and engaging training to facilitate rapid recovery. VR is a group of interventions that provides a virtual environment through computer-generated means and delivers real-time feedback on patients' performance. VR is effective in improving the motor functional status in subacute stroke patients, and this is a better option to use as adjunct therapy with conventional treatment (Q.-C. Peng et al., 2021). VR games include the Xbox Kinect and the Nintendo Wii; it is reported that Wii Sports and game parties are used in training. The Xbox games include "Good View Hunting and Hong Kong Chef", which require moving hands to pick objects for a good score. They used to move their shoulders to improve the range of motion (Leong et al., 2022b). The role of VR in cognitive rehab shows that VR, along with conventional training, improves the cognitive level, attention deficit, executive function, and reduces depressive moods in chronic stroke patients (Gao et al., 2021a).

Immersive VR, through head-mounted displays, and non-immersive VR, through screens or cameras, are used for task-oriented training. It is divided into two types: serious games are developed for specific purposes in conjunction with clinical goals, and there is progression in game complexity. Commercial off-the-Shelf (COTS) games are designed for home-based and clinical settings to facilitate ease of access and engagement. The VR gained increased attention after COVID-19 and was more widely implemented across different settings for stroke rehabilitation. The Fully Immersive Virtual Reality (FIVR) has a significant impact on gait, postural control, and upper-limb proprioception. FIVR has engaging effects on motor restoration, functional status, and quality of life in patients with stroke. Immersive VR has beneficial effects on the functional status of the upper and lower limbs in stroke (Demeco et al., 2023). In the last few years, Video Games (VG) have shifted from passive to active modes, where players participate physically, moving their bodies or parts of their bodies to play. These games require higher levels of physical activity and mental engagement. In addition to conventional intervention, VG is beneficial in promoting physical activity in neurological conditions. However, as these are not specific for clinical purposes, they have multiple limitations. Preferably, clinically oriented, customised games should be developed and used to take full advantage (Bonnechère et al., 2016).

The American Physical Therapy Association reported that VR can provide customised rehabilitation for stroke, Parkinson's, and multiple sclerosis. Canadian stroke best practices also endorsed that VR can be used as adjunct therapy and recommended it whenever no direct care is available. However, the UK National Institute for Health and Care Excellence (NICE) stroke guidelines noted that VR games are costly and uncommon in telerehabilitation. As stroke pathways are expected to shorten and inpatient rehab has been reduced, VR technology offers certain benefits for functional outcomes in supervised clinical settings, particularly for upper-limb, gait, and balance impairments. COTS have more engaging effects, and overall adherence is good. The dosage for VR is highly intensive, which means more than 20 hours, more than 4 times per week, and more than 60 minutes daily intervention is better to improve the overall functional and activity level in chronic stroke (Gao et al., 2021b).

The AR system provides a broader range of high-intensity training options to promote functional recovery and motor function after stroke. This device provided motivation and engagement for participants' upper- and lower-limb function. AR includes the Fruit Ninja game, in which participants must perform various quick movements. They offered to perform real-life functions and provide somatosensory and visual feedback. VAMR (Virtual Augmented Mixed Reality) has been growing rapidly recently and is considered a beneficial and effective strategy. Reh@task, MoU-Rehab, Smart Gloves, Smart Board, Reh@City, Lokomot with VR, CAREN, and Intersense Reality System are used for daily task activities. It has promising effects on upper-limb function and daily activities, but not on upper-limb fine-motor skills, as measured by the Wolf Motor Function Test. The dosage is once to 7 times a week, with sessions lasting 30 to 120 minutes. It is also reported that AR is effective for balance improvement and is superior to home-based training in a hospital setting (Gao et al., 2021b). 409The future of AR should be integrated with artificial intelligence, encompass a broader range of HMDs, provide quick feedback, and support the gradual progression of tasks for each stroke patient (Jia et al., 2025). The system should be low-cost, user-friendly, accessible, and portable to promote motivation and wider utilisation of such devices. It is also suggested that real-time feedback and progress reviews be provided to stroke rehab professionals to support appropriate clinical decision-making (Phan et al., 2022).

Brain–Computer Interface (BCI) has three types commonly used in stroke rehabilitation: first, Motor Imagery (MI), in which patients imagine moving the hand without physically moving it. Second, Intention Movement (IM), in which the patient physically tries to move the hand; it is also called a motor-attempt-based BCI. The third type is action observation BCI, in which the patient observes an action before attempting it. In stroke rehab, BCI is mainly achieved via EEG, a non-invasive approach, and visual feedback is also provided to enhance performance. In addition to mental imagery training, BCI has a significant impact on stroke rehabilitation, especially in severe impairment. BCI intervention has a positive impact on restoring upper-limb function in subacute stroke patients and improves safety and quality of life (J. Liu et al., 2025b). Four weeks of BCI training, along with other conventional training and Motor imagery training, have better outcomes for upper-extremity function after stroke and are considered a safe approach to training (Q.-C. Peng et al., 2021). BCI has a promising effect on upper-limb recovery and ADLs. BCI, along with Brain Machine Interface, FES, or visual feedback, improves upper-limb function post-stroke compared with robotics alone (Xie et al., 2022).

Wearable Sensors (WS) devices are widely used in stroke rehab for precise assessment and training, both in clinical settings and home-based sessions. The assessment includes upper-extremity motion, motor impairment, activity limitation, and real-world activities. These assessments facilitate the training protocol through sensory cues and objective-oriented performance feedback at home. This feedback provides specific data on activity level and adherence to home-based training. This device offers objective assessment and automated intervention to prevent stroke impairments and promote stroke recovery. High-intensity stroke training is recommended for quick recovery and motor recovery (Kim et al., 2022b). However, intensive long-term training requires many therapists, multiple stroke centres, a high budget for staff salaries, and high hospitalisation costs, among other direct and indirect costs. Home-based training is considered a potential solution, and self-directed training can reduce healthcare costs. Patients and caregivers manage self-directed training without therapist supervision (Toh et al., 2023).

WS devices are suitable options for home-based training in self-directed exercises, and this approach is cost-effective for longer-term stroke rehabilitation. These are low-cost, user-friendly, portable, flexible, and easy to wear. These devices measure motion, posture, and kinematics and provide real-time motion-correction feedback to assist movements. Few WS provide real-time feedback through visual, auditory, or tactile modalities to enhance motor relearning. This feedback argues for posture and movement corrections during task performance. Intrinsic feedback from propriocep-tion is impaired in stroke patients, while therapist-oriented feedback is always time-consuming and not available to correct the task performance. However, this device-oriented feedback is available to patients at home for self-directed learning and enhanced functional performance. Monitoring and progress monitoring are good for training adherence. WS is an effective home-based approach for stroke rehabilitation of upper and lower-limb function (Toh et al., 2023).

Telerehabilitation provides a digital platform for engaging patients in remote areas. The American Heart Association guidelines and Canadian stroke best prac-tices state that stroke rehabilitation should be provided for at least three hours per day, five days per week, during inpatient care. This intensive stroke training is a serious challenge for the healthcare system to fulfil the guidelines due to numerous barriers and limitations, including the limited staff, other patient health issues, and a high number of patients. Mobile applications (apps) are innovative, emerging technolo-gies that are becoming increasingly popular due to their accessibility, mobility, and multifaceted functionality, such as progress monitoring and video-based guidelines. Mobile apps are effective in stroke rehabilitation, as they facilitate finger dexterity, adherence programs, home-based upper-extremity training, and mirror therapy for facial palsy. The four mobile apps include therapy apps, rehab videos, reminders, and rehab videos with reminders. These apps are designed to enhance the repeti-tion of specific activities through game-based practice. This repetitive task-specific practice improves motor relearning and motor recovery of function. The mobile apps improve stroke recovery when used in addition to face-to-face motor function training, aphasia training, and adherence to training. These apps provide additional training during inpatient care and meet the required duration according to standard guidelines (Szeto et al., 2023).

mHealth apps are well-being services delivered via mobile apps that provide information and communication. This technology has gained interest because it is believed to support rehabilitation goals, enhance self-directed management, and promote high levels of adherence at home and during exercise. The mhealth apps targeted 32% of upper-limb function, 26% of prevention and medical management, and 24% of exercises or mobility.mHealth apps in stroke rehabilitation focus on phys-ical, cognitive, and speech therapy. Three categories of games are available: gaming apps, exercise prescription apps, and monitoring apps. The gaming app includes Upper Extremity (MoU-Rehab, unnamed finger-training app, FINDEX, Tap it, and ARMStroke) and postural control (unnamed app). The exercise prescription apps are relevant to mobility (CARE4STROKE) and motor function (9zest Stroke Rehab and Farmalarm), while the monitoring apps target physical activity (FitlabVR and

STARFISH). The Mhelath apps, with their physical training section, promote function and activity levels in stroke rehab and are recommended as additional support in post-stroke rehab (Rintala et al., 2023).

2.7.2 Parkinson's Disease

Parkinson's disease (PD) is a neurodegenerative progressive disorder due to loss of dopamine in the substantia nigra and accumulation of Lewy bodies. Parkinson's disease (PD) is a progressive disorder and has doubled globally over the past two decades. Parkinson's disease has shifted from 19th to 14th in DALYs with 128% increase in 2019. The rising statistics in NDs are alarming for healthcare systems (Ding et al., 2022c). The clinical motor symptoms of PD include postural instability, rigidity, tremors, and gait abnormality, while nonmotor symptoms include sleep disorder, cognitive impairment, mood disorders, and autonomic dysfunction. These clinical symptoms lead to impaired mobility and a decreased quality of life for patients. Pharmacological treatment is available for PD. However, PD has some mobility issues leading to poor activity, daily activities, and participation. Therefore, rehabilitation is a significant part of management (Lazzarini et al., 2024).

Robotic-assisted technology focuses on motor relearning, providing support and assistance for intensive, repetitive, and task-oriented training to enhance neural plasticity through continuous stimulation. The literature provides uncertain evidence regarding the effectiveness of robotics for PD motor, balance, and upper-limb outcomes compared with other interventions. The literature reports that RAT effectiveness is still under review; it shows better motor outcomes and improves balance compared to conventional physical therapy (Lazzarini et al., 2024). RAT improves motor function, balance confidence, and fatigue levels, but there is no evidence of improvement in upper-limb function. It is also suggested that RAT with virtual reality is more effective in PD. Robot-assisted gait training benefits people with PD; however, it is not considered superior to other interventions. RAT is beneficial for patients with severe disease and symptoms, improving their functional levels. (Alwardat et al., 2018; Carmignano et al., 2022; Tao et al., 2024).

Virtual Reality (VR) plays an engaging role in PD rehabilitation. Postural control and gait are common impairments in PD, and balance instability is evident in its early stages. Balance training programs are effective in reducing fall risk and improving functional status and quality of life. The VR rehabilitation-specific system is a customised platform that includes BioFlex-FP, IREX, and NIRVANA for training across different functional domains. Non-customised VR systems, including the Xbox Kinect and the Nintendo Wii, are commonly used in balance training for PD (Sarasso et al., 2022a). VR has been used recently in clinical and home settings for balance and gait training for PD. VR provides sensory, tactile, and visual stimulation to enhance neural circuit engagement across multiple senses. The non-immersive VR includes Xbox 360 Kinect adventure games, Nintendo Wii, and the monitor-based game "Pull the Body." In contrast, the immersive VR includes the VR Box headset,

NIRVANA, and the C-Mill treadmill with VR projection. VR-based training shows improved balance and gait as an adjunct to conventional training.

However, it remains unclear whether VR training is superior to conventional interventions, despite adherence and motivation being key aspects of VR training. There is still a gap regarding the standard type of training, variations in game formats, training duration, and specific customisation for different stages of the disease (Fernandes et al., 2025). VR is effective for improving balance, as it integrates weight shifting, enhances proprioception, and supports balance maintenance during gameplay. The virtual environment offers a stress-free environment for certain activities and an engaging atmosphere. It provides high confidence for managing real-world tasks. VR also has certain positive effects on cognitive impairment (Kwon et al., 2023). The training ranges from 2–5 days per week to 4–12 weeks, with sessions lasting 20–60 minutes. There is significant improvement in balance and mobility, but no effect on walking speed has been reported. However, follow-up at 1–3 months has shown no maintenance of recovery despite established effects of VR on motor relearning (Sarasso et al., 2022b).

Brain–Computer Interface (BCI) is a technology-based approach for decoding abnormal brain activity and stimulating normal movement parameters. Multiple pathophysiological and metabolic changes occur in the brains of patients with PD. BCI has the potential to activate neural plasticity in the brain and promote normal movement patterns by focusing on beta-band activity, phase-amplitude coupling, and alpha-band dynamics observed in EEG. Different BCI combinations are used in PD, including EEG-BCI neurofeedback through motor imagery for motor and cognitive training, or EEG-BCI with attention control only for cognitive training. Functional MRI-BCI neurofeedback through motor imagery for hand training and BCI-DBS-controlled adaptive unilateral and bilateral for overall motor training. With certain limitations, EEG-BCI is a promising and transformative approach for monitoring and adaptive, personalised intervention for PD (Ortega-Robles et al., 2025). Motor Imagery-based BCI enhances the performance in PD and improves the quality of life. Deep Brain Stimulation (DBS) is an evidence-based intervention for PD, while BCI is an emerging technology for addressing motor disturbances in PD. Advanced DBS systems share some similarities with BCI systems, as they monitor brain activity in real-time and provide personalised interventions. Further clinical research will examine the details and long-term effects of PD (H. Zhang et al., 2024).

2.7.3 Spinal Cord Injury (SCI)

The incidence of SCI is 13–220 cases per million people worldwide. SCI is a highly devastating injury which affects motor function, sensory function, and bowel and bladder control. There are certain consequences or complications of SCI, including cardiopulmonary, respiratory, gastrointestinal complications, muscle weakness, osteoporosis, and pressure sores, leading to decreased life expectancy. These injuries directly affect motor function, activity level, and participation, leading to a decline in quality of life and an increasing socioeconomic burden on society. The prognosis of recovery is limited, and patients face long-term physical and cognitive impairment. Emerging technologies play a significant role in the rehabilitation of patients with SCI by providing support and assistance with functional tasks and mobility (Guo et al., 2025; S. Liu et al., 2025).

Robot-Assisted Technology in SCI provides a supportive unit that facilitates movement activation and enhances independence through repetitive task practice. Robot-assisted training aids weight-bearing and reduces exertion intensity post-SCI. The outcome of SCI training depends mainly on the nature of the injury, its location, its duration, and the timing of rehabilitation. In complete SCI, exoskeletons enhance dexterity, manage pain, and reduce spasticity, but offer limited improvement in independent walking. The HLA (Hybrid Assistive Limb) exoskeleton, along with a treadmill with high-repetition and active stepping feedback, increases brain network connectivity in partial SCI. Exoskeleton robot training in SCI is effective in improving motor function, including walking, distance, speed, and lower extremity function. It is effective within six months; beyond that, results are unclear. Medium-intensity training (1000–2000 min) has better outcomes than mid- and high-intensity training regimes (Gao et al., 2021a). The literature suggests that walking training is superior to non-walking training for motor recovery and neuromuscular performance in SCI. Robotic Exoskeleton Gait Training (REGT) is an innovative device for SCI rehabilitation, including Ekso Bionics, Cyberdyne, and Rex Bionics. These devices assist patients in standing upright and walking. These devices provide sensory integration to the brain and promote functional generator patterns in the spinal cord. Robotic Exoskeleton gait training is an effective strategy for improving walking balance, functional scores, breathing function, and lower-limb strength compared with conventional training. However, the impact on walking speed and distance post-injury is limited (Gao et al., 2021b). Robot-Assisted Gait Training (RAGT) is also beneficial for improving timed-up-and-go test performance in SCI and overall mobility. Still, it has no significant effect on gait distance, speed, leg length, or 6-MWT (Bin et al., 2023). The RAGT has remarkable improvements in activities of daily living, muscular strength, walking ability in SCI, and effectiveness that lasts>2 months among subacute patients (Bin et al., 2023).

Virtual Reality (VR) technologies have been widely used, including the Xbox Kinect, GRAIL training, the Sony PlayStation 2, the EyeToy, and the Nintendo Wii Fit, to promote active participation and engaging activities. The games and exercises include soccer games, snowboard simulation, and weight-shifting stability to improve balance, mobility, and functional status. VR is effective in SCI to improve balance, mobility, and functional outcomes by promoting neural plasticity, motor relearning, and strengthening central nervous system connections. However, the type of VR, patient selection criteria, and other related biases need to be explored (Mohamad et al., 2020). Upper-limb motor function, shoulder strength, and strengthening of the weaker side are outcomes of VR-assisted rehabilitation. The dosage includes 30–60 minutes per session, 2–5 sessions per week, and for 3–7 weeks (De Miguel-Rubio et al., 2022a). The literature is still controversial regarding the effectiveness of VR in SCI. VR improves walking ability, motor function, and balance outcomes in

SCI; however, it has no significant effects on lower-limb function or daily activities (De Miguel-Rubio et al., 2022b).

2.7.4 Traumatic Brain Injury (TBI)

TBI is a non-progressive brain injury caused by external forces. TBI is one of the leading causes of disability in moderate to severe cases and affects 64 million people annually. The impairments of TBI include physical, emotional, speech, and cognitive effects that affect activity levels and reduce quality of life. It also affects higher mental functions, including memory, attention, and behaviour, especially damaging the frontal and temporal areas. These impairments contribute to the limitation of daily activities, increase dependence and cause long-term disability (Bonanno et al., 2022; Shen et al., 2025). The two cognitive rehabilitation approaches commonly used are restorative (training impaired functions) and compensatory (strategy training and external aids). Emerging technologies promote early recovery, neural plasticity, and functional regain through feedback-oriented, task-specific, and highly intensive training programs (Kaurani et al., 2024a).

Robot-Assisted Technology assists with hip, knee, and ankle movements to generate muscle torque for postural stability. Ekso, Realwalk, and HAL are rigid robotic devices used to maintain upright posture while standing in patients with TBI. Robotic devices are useful tools for providing repetitive movement and early mobility in severe cases of TBI (Karunakaran et al., 2023). These devices increase endurance, gait symmetry, speed, and mobility. Additional impact on cognitive function when used as an adjunct to AI or VR models. Robotics are used to increase the number of repetitions, dosage, and active engagement, which support functional and motor recovery. Robotic gait training, approximately 30–60 minutes per session for 6–12 weeks, seems effective in achieving the expected goals (Shen et al., 2025).

Virtual Reality (VR) improves balance, coordination and overall postural control in TBI. It also improves cognitive function, including higher-level functions such as memory and attention, in patients. It also enhances motivation and emotional stability and reduces anxiety. An almost 30% improvement in higher mental function was reported compared with conventional. VR gaming improves executive functioning and working memory; there is limited evidence on its effects on Attention-Deficit/Hyperactivity Disorder (ADHD). VR games include NeuroDRIVE for military patients, for driving training and cognitive and behavioural modification, with improved memory and selective attention during skill performance. VMall and BTS-Nirvana are effective games for improving ADLs, cognitive functioning, and visuospatial function during daily activities (Bonanno et al., 2022). Cognitive games include virtual supermarkets and sequencing and planning tasks; motor games include gait with cognitive challenges; and immersive head-mounted devices are used for training. Overall, VR games improve cognition, executive function, attention, working memory, adherence, motivation, and task planning. Immersive VR has a stronger impact on emotional and cognitive enhancement than non-immersive VR games. The dosage includes 30–60 minutes per session and 2–5 sessions per week for at least 4–8 weeks (M. Zhang et al., 2025).

Brain–computer interfaces (BCIs) enable control and motor initiation in severe conditions. BCI enhances function through structural reorganisation and functional recovery by linking neural signals to external devices or feedback. BCI promotes function through motor imagery training by detecting brain signals (EEG) and converting them into feedback for functional electrical stimulation (FES). BCI enhances upper-limb function, motor performance, and coordination in subacute and chronic patients with TBI (Kaurani et al., 2024b).

Telerehabilitation (TR) is a beneficial approach to providing accessible, cost-effective care for patients. TBI rehabilitation requires long-term care with frequent follow-up and monitoring of progress. TR is applicable via video calls, telephone, messaging, email, mobile apps, and wearable sensors in remote locations to provide access to training. The outcomes are better in terms of function and cognition, emotional well-being, and quality of life, and they promote adherence to long-term training sessions and reduce caregiver burden. Tele-rehab approach improves problem-solving skills, coping strategies, anxiety and reduces depression. The session should be 30–60 minutes, weekly or biweekly, for 6 weeks to 9 months, to address long-term disability and functioning (Avramovic et al., 2023).

Non-Invasive Brain Stimulation (NIBS), including transcranial direct current stimulation (tDCS) and repetitive transcranial magnetic stimulation (rTMS), is a non-invasive approach for stimulating brain areas and enhancing neural plasticity. These devices are effective when used in conjunction with traditional rehabilitation. It enhances memory and attention during cognitive function and improves divided attention by normalising abnormal hyperactivity in temporal and frontal areas. Improvement in depression, fatigue, and motor activities have been observed when NIBS is combined with traditional task-specific training. The NIBS approach is effective in targeting specific functions, such as cognitive, motor, or mood, in TBI. The long-term effectiveness, dosage, and stimulation parameters still need further research to develop a protocol framework (Bonanno et al., 2022).

2.7.5 Multiple Sclerosis (MS)

MS is a progressive, chronic, neuroinflammatory autoimmune, and demyelinating disease of the central nervous system. The incidence of MS is 1.2 per 100, 000 people across 75 countries. The symptoms of MS include motor weakness, ataxia, sensory impairments, bowel and bladder dysfunction, and abnormal gait patterns. Various pharmacological treatments are available in literature. Still, as there is no specific cure for MS, rehabilitation is the appropriate option to improve motor recovery and enhance functional status (Herrera-Rojas et al., 2025; F.-A. Yang et al., 2023). Digital technology is a part of comprehensive MS rehabilitation and assessment, including exergaming, robotic-assisted exercises, and AI-driven tools. These tools are better options for improving motor and cognitive rehabilitation, reducing monitoring time, and improving training adherence (De Angelis et al., 2021).

Robot-Assisted Technology includes grounded exoskeletons that are used in combination with traditional rehabilitation to improve gait at normal speed through repeated practice over a longer duration. The effects of grounded exoskeletons are positive for the short term and are considered a better option for MS rehabilitation (J. Yang et al., 2023). Robot-assisted gait training improves balance and gait speed, but overall walking performance remains the same compared to traditional training. Robot-assisted training is considered safe, secure, and without any adverse effects. It also reduces fatigue and spasticity and improves mobility and quality of life in people with MS. In moderate-to-severe MS, robot-assisted upper-limb training augmented with serious games improves cognitive and motor function through repetitive, highly intensive therapy. Feedback in robot-assisted training is strongly linked with emotional, cognitive, and motivational aspects of people with MS. The dosage includes 20–60-minute sessions, 3–5 per week for 4–8 weeks, but it varies across different levels of severity and other factors associated with MS (Facciorusso et al., 2024).

Telerehabilitation (TR) is a suitable option for MS to provide care at home or in remote areas, reducing travel time and distance. It also reduces the daily cost of travel to rehabilitation centres for long-term care for MS. TR has a significant impact on health-related quality of life, motor recovery, neural plasticity, treatment adherence, and positive feedback. TR promotes a multidisciplinary team approach in MS to improve prognosis. TR is beneficial for people with geographical, financial, and accessibility disadvantages and provides uniform management options for all patients with MS (Herrera-Rojas et al., 2025). Mobile applications are useful tools for tracking and monitoring changes. Fatigue is one of the most devastating symptoms in people with MS. Mobile apps are used to manage fatigue through self-management, lifestyle modification, real-time feedback, and active engagement support. Mobile apps have a significant impact on cognitive function, motor recovery, quality of life, and education for people with MS. Mobile apps should be utilised for 30 minutes over 8 weeks to engage patients in meaningful outcomes (Facciorusso et al., 2024).

Virtual Reality (VR) enhances balance and postural control in MS as compared to traditional training strategies. There is no significant impact of VR on walking speed and mobility. Kinect- or Wii-based VR also improves balance and reduces fear of falling. VR training provides an engaging environment that enhances cognitive function, including attention, higher mental function, memory, and visuospatial skills. Literature is limited, but it supports the idea that VR also improves the overall quality of life and reduces fatigue severity in people with MS. The suggested VR training ranges from 20 to 40 sessions, 2–5 days a week, to improve balance and functional mobility (Kamari et al., 2024).

Non-Invasive Brain Stimulation (NIBS) tools include transcranial direct current stimulation (tDCS) and repetitive transcranial magnetic stimulation (rTMS). These tools are useful for reducing spasticity in MS and are considered safe. The protocol includes short sessions of 3–5 for 15–30 minutes for 2–4 weeks, but no standard dosage has been set for its application. The tDCS at 1.5 and 2 mA intensities shows better outcomes for cognitive function, reduces fatigue at low physical disability levels, and has no effect at high physical disability levels. The rTMS has a clinical impact on the reduction of spasticity associated with MS (Kan et al., 2022).

2.7.6 *Amyotrophic Lateral Sclerosis (ALS)*

ALS is a progressive neurodegenerative motor neuron disease with symptoms of both upper and lower motor neurons. The symptoms include motor and extrapyramidal, leading to death after 3–5 years of onset. The motor dysfunction includes spasticity, muscular weakness, dysphasia, impaired gait, limited mobility, and cognitive impairment. The most devastating symptoms include the loss of respiratory muscle weakness and other pulmonary complications, which can lead to death in most cases. The incidence is 2 out of 100,000, and the prevalence is 3–8 per 100,000 worldwide. The intervention is multidisciplinary, focusing on all aspects of patients, and it promotes quality of life and supports survival in patients with ALS. External assistance and support through emerging technologies play a significant role in maintaining mobility and enhancing overall functional level (Calderone et al., 2026a).

Robot-Assisted Technology includes mortised and robot-assisted systems that provide support for mobility and functional task performance. They promote repetitive tasks and optimise patient requirements to achieve specific goals. Robot-assisted training with the HAL wearable exoskeleton promotes gait patterns and enhances walking capacity. The recommended dosage is 20–40-minute sessions, 9–10 sessions over 4 weeks, for meaningful results. This training has no side effects and is considered a safe protocol for patients with ALS (Pugliese et al., 2022).

Wearable Devices include soft robotics that provide personalised assistance to promote mobility, improve gait performance, facilitate movement, and reduce physical exertion. It assists the gait pattern through the hip, knee, and ankle joints and facilitates a smooth gait cycle. This device is lightweight, flexible and enhances motor control as compared to a rigid exoskeleton. It also reduces metabolic cost and energy consumption during assisted walking, thereby improving walking efficiency. Utilisation is better in the short term and enhances walking performance soon after device use (Calderone et al., 2026b). Wearable soft robotics for upper-limb function, a short, fabric inflatable pneumatic actuator around the shoulder and axilla. The sensors detect voluntary muscular activity signals and adjust the robot accordingly, as they are safe, comfortable, and lightweight. This training provides short-term functional improvement, reduces muscle fatigue, improves grip strength, and yields a good user experience (Proietti et al., 2023).

Telerehabilitation (TR) provides care at home with monitoring support systems, including support for breathing and dysphasia. This approach provides an accessible and remote training program, as progressive functional decline is characteristic of ALS. Telerehabilitation through video conferencing and wearable monitoring (vital signs) improves the respiratory function, inspiratory muscle strength, activities of daily life, and overall quality of life. TR provides counselling and education to family and caregivers to reduce anxiety, improve coping strategies, and enhance the quality of life of patients and family members. TR reduces travel costs and the frequency of hospital visits, especially in the mid- to late-stage of disease. TR is an accessible, feasible, and appropriate strategy for home-based rehabilitation and for improving motor function and activity levels compared to other exercise types (Kato et al., 2025).

2.8 Limitations and Challenges of Emerging Technologies

There are certain limitations of technologies in the rehabilitation of neurological conditions. These limitations have different impacts on patients and therapists according to the nature of the challenge. The limitations related to patients, technologies, and the organisational environment directly and indirectly affect intervention compliance and outcomes.

2.8.1 Patient-Related Challenges

The challenge of emerging technologies in neurorehabilitation is directly linked to patients' associated factors. Low digital literacy, especially among older adults, is a major challenge in LMICs. Patients experience discomfort when using VR, wearable sensors, or robotic devices, which affects intervention adherence. Moderate to severe cognitive impairments in NDs impair the proper understanding of instructions and the ability to follow commands when interacting with technological devices. The severity of disease is a critical challenge for the effective implementation of emerging technologies. Severe motor dysfunction, significant spasticity, end-stage rigidity, and complete dependency hinder the proper utilisation of technologies. Fatigue and pain also affect intensive, repetitive training programs. Depression and mood variation lead to low motivation to engage in and active participation in technologically oriented training, as this is required for effective outcomes (Jarvis et al., 2024; Nizamis et al., 2021).

2.8.2 Technology-Related Challenges

Technology has maximum potential to support the rehabilitation of patients with NDs and enable them to become independent individuals. However, there are a few challenges associated with technologies, including installation processes, complex SOPs, maintenance issues, calibration procedures, software upgrades, interface restrictions, and user-friendliness, which affect overall progress. Hardware and software limitations disrupt the continuity of intervention, and internet connectivity issues reduce the efficacy of telerehabilitation programs (Alt Murphy et al., 2024b; Nizamis et al., 2021).

2.8.3 Organisation-Related Challenges

The requirements for infrastructure and resources are major challenges for implementing emerging technology-based intervention programs in clinical settings, especially in LMICs. Limited availability of technology and devices restricts exposure to all patients, and only some specific patients benefit from these interventions. Poor internet connectivity and a lack of trained IT staff reduce the appropriate implementation of technology-driven programs, especially in remote areas. Lack of policy and financial constraints to adopt these high technologies is a serious challenge, especially in LMICs. Limited community centres and a lack of a standardised implementation framework lead to inconsistencies in the long-term sustainability and service delivery of emerging technologies (Alt Murphy et al., 2024b).

2.8.4 Future Directions in Technological Assisted Neurorehabilitation

There is rapid technological advancement to fulfil the needs and demands of the community. For future work, a few suggestions to improve the effectiveness of neurorehabilitation include the development of highly personalised systems, task difficulty, complexity, feedback, and progression based on cognitive, physical, or psychological level. The integration of an AI-driven system to deliver a more precise solution and the engagement of caregivers, clinicians, and patients in collaboration to improve resource utilisation. The new models should align with the principles of neural plasticity, repetition, intensity, engagement, task specificity, feedback, salience, and transfer. Digital literacy is a key factor in the effective implementation of technology; it should be included in the curriculum to foster better understanding and embed technology education. There is also a need to train the staff and the patient community through an awareness campaign. A hybrid technology model should be integrated to involve therapists for guidance and precise assessment. An expansion of digital home-based care, with minimal resources and a user-friendly interface, should be introduced to improve long-term care. The development of cost-effective, low-burden, low-budget devices for low-income countries should focus on serving large populations. Policies should be prepared and approved to support emerging technologies in insurance, reimbursement, and data security. The ethical aspects should also be highlighted, and an ethical framework should be devised for implementation in the true spirit. Emotional and human interaction is incorporated into the emerging technology to enhance the recovery and confidence of therapists and patients (Cano-de-la-Cuerda et al., 2024).

2.9 Conclusion

Emerging technologies are transforming the shape of neurological rehabilitation of physical disabilities. These technologies address the major constraints of traditional interventions: intensity, dosage, repetition, individualised interventions, accessibility, monitoring, and feedback. Neurological disorders include stroke, SCI, Parkinson's, TBI, multiple sclerosis, and motor neuron diseases, which have been rapidly increasing in global burden. There is an urgent need to provide accessible, evidence-based interventions for this growing population to reduce the social and economic burden. The emerging technologies include robot-assisted training, virtual reality, augmented reality, BCI, wearable sensors, telerehabilitation, mobile apps, and AI-driven systems to provide engaging interventions for neurological disorders. These technologies provide real-time feedback, goal-specific training, personalised intervention, an accessible, highly intensive program, and reduce the physical burden of staff. However, these technologies are not a complete replacement for training, as human interaction is vital for long-term, effective recovery. Other challenges include high cost, lack of training, ethical considerations, and infrastructure requirements, all of which are barriers to effective utilisation. Future directions include a personalised intervention plan supported by an AI-driven platform, as well as cost-effective technologies and policy implementation to ensure long-term adoption.

In conclusion, emerging technologies represent a paradigm shift from traditional to interactive and engaging training tools in neurological rehabilitation. These provide an enriching environment that improves functional recovery and enhances the quality of life of people with neurological disabilities. The incorporation of emerging technologies into the intervention plan improves the quality of life and enables individuals to become independent and productive members of the community.

References

Alt Murphy, M., Pradhan, S., Levin, M. F., & Hancock, N. J. (2024a). Uptake of Technology for Neurorehabilitation in clinical practice: A scoping review. *Physical Therapy, 104*(2), pzad140. https://doi.org/10.1093/ptj/pzad140

Alt Murphy, M., Pradhan, S., Levin, M. F., & Hancock, N. J. (2024b). Uptake of Technology for Neurorehabilitation in clinical practice: A scoping review. *Physical Therapy, 104*(2), pzad140. https://doi.org/10.1093/ptj/pzad140

Alwardat, M., Etoom, M., Al Dajah, S., Schirinzi, T., Di Lazzaro, G., Sinibaldi Salimei, P., Biagio Mercuri, N., & Pisani, A. (2018). Effectiveness of robot-assisted gait training on motor impairments in people with Parkinson's disease: A systematic review and meta-analysis. *International*

Journal of Rehabilitation Research, 41(4), 287. https://doi.org/10.1097/MRR.000000000000 0312

Avramovic, P., Rietdijk, R., Attard, M., Kenny, B., Power, E., & Togher, L. (2023). Cognitive and behavioural digital health interventions for people with traumatic brain injury and their caregivers: A systematic review. *Journal of Neurotrauma, 40*(3–4), 159–194.

Banyai, A. D., & Brişan, C. (2024a). Robotics in physical rehabilitation: Systematic review. *Healthcare, 12*(17), 1720. https://doi.org/10.3390/healthcare12171720

Banyai, A. D., & Brişan, C. (2024b). Robotics in physical rehabilitation: Systematic review. *Healthcare, 12*(17), 1720. https://doi.org/10.3390/healthcare12171720

Baronchelli, F., Zucchella, C., Serrao, M., Intiso, D., & Bartolo, M. (2021). The effect of robotic assisted gait training with Lokomat® on balance control after stroke: Systematic review and meta-analysis. *Frontiers in Neurology, 12*, 661815. https://doi.org/10.3389/fneur.2021.661815

Barrett, S., Howlett, O., Lal, N., & McKinstry, C. (2024). Telehealth-delivered allied health interventions: A rapid umbrella review of systematic reviews. *Telemedicine Journal and E-Health: The Official Journal of the American Telemedicine Association, 30*(6), e1649–e1666. https://doi.org/10.1089/tmj.2023.0546

Bhujel, S., & Hasan, S. (2023). A comparative study of end-effector and exoskeleton type rehabilitation robots in human upper extremity rehabilitation. *Human-Intelligent Systems Integration, 5*(1–2), 11–42. https://doi.org/10.1007/s42454-023-00048-y

Bin, L., Wang, X., Jiatong, H., Donghua, F., Qiang, W., Yingchao, S., Yiming, M., & Yong, M. (2023). The effect of robot-assisted gait training for patients with spinal cord injury: A systematic review and meta-analysis. *Frontiers in Neuroscience, 17*, 1252651. https://doi.org/10.3389/fnins.2023.1252651

Bonanno, M., De Luca, R., De Nunzio, A. M., Quartarone, A., & Calabrò, R. S. (2022). Innovative Technologies in the Neurorehabilitation of traumatic brain injury: A systematic review. *Brain Sciences, 12*(12), 1678. https://doi.org/10.3390/brainsci12121678

Bonnechère, B., Jansen, B., Omelina, L., & Van Sint Jan, S. (2016). The use of commercial video games in rehabilitation: A systematic review. *International Journal of Rehabilitation Research. Internationale Zeitschrift Fur Rehabilitationsforschung. Revue Internationale De Recherches De Readaptation, 39*(4), 277–290. https://doi.org/10.1097/MRR.0000000000000190

Calafiore, D., Negrini, F., Tottoli, N., Ferraro, F., Ozyemisci-Taskiran, O., & de Sire, A. (2022). Efficacy of robotic exoskeleton for gait rehabilitation in patients with subacute stroke: A systematic review. *European Journal of Physical and Rehabilitation Medicine, 58*(1), 1–8. https://doi.org/10.23736/S1973-9087.21.06846-5

Calderone, A., Latella, D., De Luca, R., Gangemi, A., Impellizzeri, F., De Pasquale, P., Corallo, F., Manuli, A., Quartarone, A., Portaro, S., & Calabrò, R. S. (2026a). Virtual horizons: Enhancing rehabilitation of neuromuscular diseases through virtual reality and gamification. *Journal of Neuromuscular Diseases, 13*(1), 94–108. https://doi.org/10.1177/22143602241311194

Calderone, A., Latella, D., De Luca, R., Gangemi, A., Impellizzeri, F., De Pasquale, P., Corallo, F., Manuli, A., Quartarone, A., Portaro, S., & Calabrò, R. S. (2026b). Virtual horizons: Enhancing rehabilitation of neuromuscular diseases through virtual reality and gamification. *Journal of Neuromuscular Diseases, 13*(1), 94–108. https://doi.org/10.1177/22143602241311194

Cano-de-la-Cuerda, R., Blázquez-Fernández, A., Marcos-Antón, S., Sánchez-Herrera-Baeza, P., Fernández-González, P., Collado-Vázquez, S., Jiménez-Antona, C., & Laguarta-Val, S. (2024). Economic cost of rehabilitation with robotic and virtual reality Systems in People with neurological disorders: A systematic review. *Journal of Clinical Medicine, 13*(6), 1531. https://doi.org/10.3390/jcm13061531

Carmignano, S. M., Fundarò, C., Bonaiuti, D., Calabrò, R. S., Cassio, A., Mazzoli, D., Bizzarini, E., Campanini, I., Cerulli, S., Chisari, C., Colombo, V., Dalise, S., Gazzotti, V., Mazzoleni, D., Mazzucchelli, M., Melegari, C., Merlo, A., Stampacchia, G., Boldrini, P., … Andrenelli, E. (2022). Robot-assisted gait training in patients with Parkinson's disease: Implications for clinical practice. A systematic review. *Neuro Rehabilitation, 51*(4), 649–663. doi:https://doi.org/10.3233/NRE-220026.

De Angelis, M., Lavorgna, L., Carotenuto, A., Petruzzo, M., Lanzillo, R., Brescia Morra, V., & Moccia, M. (2021). Digital technology in clinical trials for multiple sclerosis: Systematic review. *Journal of Clinical Medicine, 10*(11), 2328.

De Miguel-Rubio, A., Muñoz-Pérez, L., Alba-Rueda, A., Arias-Avila, M., & Rodrigues-de-Souza, D. P. (2022a). A therapeutic approach using the combined application of virtual reality with robotics for the treatment of patients with spinal cord injury: A systematic review. *International Journal of Environmental Research and Public Health, 19*(14), 8772. https://doi.org/10.3390/ijerph19148772

De Miguel-Rubio, A., Muñoz-Pérez, L., Alba-Rueda, A., Arias-Avila, M., & Rodrigues-de-Souza, D. P. (2022b). A therapeutic approach using the combined application of virtual reality with robotics for the treatment of patients with spinal cord injury: A systematic review. *International Journal of Environmental Research and Public Health, 19*(14), 8772. https://doi.org/10.3390/ijerph19148772

Del Pino, R., Díez-Cirarda, M., Ustarroz-Aguirre, I., Gonzalez-Larragan, S., Caprino, M., Busnatu, S., Gand, K., Schlieter, H., Gabilondo, I., & Gómez-Esteban, J. C. (2022). Costs and effects of telerehabilitation in neurological and cardiological diseases: A systematic review. *Frontiers in Medicine, 9*, 832229. https://doi.org/10.3389/fmed.2022.832229

Demeco, A., Zola, L., Frizziero, A., Martini, C., Palumbo, A., Foresti, R., Buccino, G., & Costantino, C. (2023). Immersive virtual reality in post-stroke rehabilitation: A systematic review. *Sensors, 23*(3), 1712. https://doi.org/10.3390/s23031712

Ding, C., Wu, Y., Chen, X., Chen, Y., Wu, Z., Lin, Z., Kang, D., Fang, W., & Chen, F. (2022a). Global, regional, and national burden and attributable risk factors of neurological disorders: The global burden of disease study 1990–2019. *Frontiers in Public Health, 10*. https://doi.org/10.3389/fpubh.2022.952161

Ding, C., Wu, Y., Chen, X., Chen, Y., Wu, Z., Lin, Z., Kang, D., Fang, W., & Chen, F. (2022b). Global, regional, and national burden and attributable risk factors of neurological disorders: The global burden of disease study 1990-2019. *Frontiers in Public Health, 10*, 952161. https://doi.org/10.3389/fpubh.2022.952161

Ding, C., Wu, Y., Chen, X., Chen, Y., Wu, Z., Lin, Z., Kang, D., Fang, W., & Chen, F. (2022c). Global, regional, and national burden and attributable risk factors of neurological disorders: The global burden of disease study 1990-2019. *Frontiers in Public Health, 10*, 952161. https://doi.org/10.3389/fpubh.2022.952161

Facciorusso, S., Malfitano, C., Giordano, M., Del Furia, M. J., Mosconi, B., Arienti, C., & Cordani, C. (2024). Effectiveness of robotic rehabilitation for gait and balance in people with multiple sclerosis: A systematic review. *Journal of Neurology, 271*(11), 7141–7155.

Feigin, V. L., Vos, T., Nichols, E., Owolabi, M. O., Carroll, W. M., Dichgans, M., Deuschl, G., Parmar, P., Brainin, M., & Murray, C. (2020a). The global burden of neurological disorders: Translating evidence into policy. *The Lancet. Neurology, 19*(3), 255–265. https://doi.org/10.1016/S1474-4422(19)30411-9

Feigin, V. L., Vos, T., Nichols, E., Owolabi, M. O., Carroll, W. M., Dichgans, M., Deuschl, G., Parmar, P., Brainin, M., & Murray, C. (2020b). The global burden of neurological disorders: Translating evidence into policy. *The Lancet. Neurology, 19*(3), 255–265. https://doi.org/10.1016/S1474-4422(19)30411-9

Fernandes, S., Oliveira, B., Sacadura, S., Rakasi, C., Furtado, I., Figueiredo, J. P., Gonçalves, R. S., & Martins, A. C. (2025). The effectiveness of virtual reality in improving balance and gait in people with Parkinson's disease: A systematic review. *Sensors, 25*(15), 4795. https://doi.org/10.3390/s25154795

Fiore, S., Battaglino, A., Sinatti, P., Sánchez-Romero, E. A., Ruiz-Rodriguez, I., Manca, M., Gargano, S., & Villafañe, J. H. (2023). The effectiveness of robotic rehabilitation for the functional recovery of the upper limb in post-stroke patients: A systematic review. *Retos, 50*, 91–101. https://doi.org/10.47197/retos.v50.99211

Gao, Y., Ma, L., Lin, C., Zhu, S., Yao, L., Fan, H., Gong, J., Yan, X., & Wang, T. (2021a). Effects of virtual reality-based intervention on cognition, motor function, mood, and activities of daily living in patients with chronic stroke: A systematic review and meta-analysis of randomized controlled trials. *Frontiers in Aging Neuroscience, 13*, 766525. https://doi.org/10.3389/fnagi. 2021.766525

Gao, Y., Ma, L., Lin, C., Zhu, S., Yao, L., Fan, H., Gong, J., Yan, X., & Wang, T. (2021b). Effects of virtual reality-based intervention on cognition, motor function, mood, and activities of daily living in patients with chronic stroke: A systematic review and meta-analysis of randomized controlled trials. *Frontiers in Aging Neuroscience, 13*, 766525. https://doi.org/10.3389/fnagi. 2021.766525

Guo, S., Yang, Y., Wang, M., Wang, D., Zhang, Y., Wang, Q., & Deng, Y. (2025). Effects of an exoskeleton robot on motor function in patients with spinal cord injuries: A systematic review and meta-analysis. *Systematic Reviews, 14*(1), 218. https://doi.org/10.1186/s13643-025-02974-1

Herrera-Rojas, A., Moreno-Molina, A., García-García, E., Molina-Rodríguez, N., & Cano-de-la-Cuerda, R. (2025). Effects of telerehabilitation platforms on quality of life in people with multiple sclerosis: A systematic review of randomized clinical trials. *NeuroSci, 6*(4), 103.

Jarvis, K., Thetford, C., Turck, E., Ogley, K., & Stockley, R. C. (2024). Understanding the barriers and facilitators of digital health technology (DHT) implementation in neurological rehabilitation: An integrative systematic review. *Health Services Insights, 17*, 11786329241229917. https://doi.org/10.1177/11786329241229917

Jia, C., Liu, X., Ning, L., & Ge, L. (2025). The effects of augmented reality on rehabilitation of stroke patients: A systematic review and meta-analysis with trial sequential analysis. *Journal of Clinical Nursing, 34*(11), 4578–4589. https://doi.org/10.1111/jocn.17730

Kamari, M., Siqueira, V., Bakare, J., & Sebastião, E. (2024). Virtual reality technology for physical and cognitive function rehabilitation in people with multiple sclerosis. *Rehabilitation Research and Practice, 2024*(1), 2020263.

Kan, R. L. D., Xu, G. X. J., Shu, K. T., Lai, F. H. Y., Kranz, G., & Kranz, G. S. (2022). Effects of non-invasive brain stimulation in multiple sclerosis: Systematic review and meta-analysis. *Therapeutic Advances in Chronic Disease, 13*, 20406223211069198. https://doi.org/10.1177/20406223211069198

Karunakaran, K. K., Pamula, S. D., Bach, C. P., Legelen, E., Saleh, S., & Nolan, K. J. (2023). Lower extremity robotic exoskeleton devices for overground ambulation recovery in acquired brain injury—A review. *Frontiers in Neurorobotics, 17*, 1014616.

Kato, N., Suzuki, R., Kaneko, H., & Horimoto, Y. (2025). Effects of telerehabilitation on physical function and activities of daily living in patients with amyotrophic lateral sclerosis: A scoping review. *Journal of Physical Therapy Science, 37*(8), 427–434. https://doi.org/10.1589/jpts.37.427

Kaurani, P., de Marchi Apolaro, A. V. M., Kunchala, K., Maini, S., Rges, H. A., Isaac, A., Lakkimsetti, M., Raake, M., & Nazir, Z. (2024a). Advances in neurorehabilitation: Strategies and outcomes for traumatic brain injury recovery. *Cureus, 16*(6), e62242.

Kaurani, P., de Marchi Apolaro, A. V. M., Kunchala, K., Maini, S., Rges, H. A., Isaac, A., Lakkimsetti, M., Raake, M., & Nazir, Z. (2024b). Advances in neurorehabilitation: Strategies and outcomes for traumatic brain injury recovery. *Cureus, 16*(6), e62242.

Kim, G. J., Parnandi, A., Eva, S., & Schambra, H. (2022a). The use of wearable sensors to assess and treat the upper extremity after stroke: A scoping review. *Disability and Rehabilitation, 44*(20), 6119–6138. https://doi.org/10.1080/09638288.2021.1957027

Kim, G. J., Parnandi, A., Eva, S., & Schambra, H. (2022b). The use of wearable sensors to assess and treat the upper extremity after stroke: A scoping review. *Disability and Rehabilitation, 44*(20), 6119–6138.

Kleim, J. A., & Jones, T. A. (2008). Principles of experience-dependent neural plasticity: Implications for rehabilitation after brain damage. *Journal of Speech, Language, and Hearing Research, 51*(1), S225–S239. https://doi.org/10.1044/1092-4388(2008/018)

Ko, M.-J., Chuang, Y.-C., Ou-Yang, L.-J., Cheng, Y.-Y., Tsai, Y.-L., & Lee, Y.-C. (2023). The application of soft robotic gloves in stroke patients: A systematic review and meta-analysis of randomized controlled trials. *Brain Sciences, 13*(6), 900. https://doi.org/10.3390/brainsci1306 0900

Kumar, J., Patel, T., Sugandh, F., Dev, J., Kumar, U., Adeeb, M., Kachhadia, M. P., Puri, P., Prachi, F., Zaman, M. U., Kumar, S., Varrassi, G., & Syed, A. R. S. (2023a). Innovative approaches and therapies to enhance neuroplasticity and promote recovery in patients with neurological disorders: A narrative review. *Cureus, 15*(7), e41914. https://doi.org/10.7759/cureus.41914

Kumar, J., Patel, T., Sugandh, F., Dev, J., Kumar, U., Adeeb, M., Kachhadia, M. P., Puri, P., Prachi, F., Zaman, M. U., Kumar, S., Varrassi, G., & Syed, A. R. S. (2023b). Innovative approaches and therapies to enhance neuroplasticity and promote recovery in patients with neurological disorders: A narrative review. *Cureus.* https://doi.org/10.7759/cureus.41914

Kwakkel, G. (2006). Impact of intensity of practice after stroke: Issues for consideration. *Disability and Rehabilitation, 28*(13–14), 823–830. https://doi.org/10.1080/09638280500534861

Kwon, S.-H., Park, J. K., & Koh, Y. H. (2023). A systematic review and meta-analysis on the effect of virtual reality-based rehabilitation for people with Parkinson's disease. *Journal of Neuroengineering and Rehabilitation, 20*(1), 94. https://doi.org/10.1186/s12984-023-01219-3

Lazzarini, S. G., Mosconi, B., Cordani, C., Arienti, C., & Cecchi, F. (2024). Effectiveness of robot-assisted training in adults with Parkinson's disease: A systematic review and meta-analysis. *Journal of Neurology, 272*(1), 22. https://doi.org/10.1007/s00415-024-12798-z

Leong, S. C., Tang, Y. M., Toh, F. M., & Fong, K. N. K. (2022a). Examining the effectiveness of virtual, augmented, and mixed reality (VAMR) therapy for upper limb recovery and activities of daily living in stroke patients: A systematic review and meta-analysis. *Journal of Neuroengineering and Rehabilitation, 19*(1), 93. https://doi.org/10.1186/s12984-022-01071-x

Leong, S. C., Tang, Y. M., Toh, F. M., & Fong, K. N. K. (2022b). Examining the effectiveness of virtual, augmented, and mixed reality (VAMR) therapy for upper limb recovery and activities of daily living in stroke patients: A systematic review and meta-analysis. *Journal of Neuroengineering and Rehabilitation, 19*(1), 93. https://doi.org/10.1186/s12984-022-01071-x

Lima, A. A., Mridha, M. F., Das, S. C., Kabir, M. M., Islam, M. R., & Watanobe, Y. (2022). A comprehensive survey on the detection, classification, and challenges of neurological disorders. *Biology, 11*(3), 469. https://doi.org/10.3390/biology11030469

Liu, J., Li, Y., Zhao, D., Zhong, L., Wang, Y., Hao, M., & Ma, J. (2025a). Efficacy and safety of brain-computer interface for stroke rehabilitation: An overview of systematic review. *Frontiers in Human Neuroscience, 19*, 1525293. https://doi.org/10.3389/fnhum.2025.1525293

Liu, J., Li, Y., Zhao, D., Zhong, L., Wang, Y., Hao, M., & Ma, J. (2025b). Efficacy and safety of brain-computer interface for stroke rehabilitation: An overview of systematic review. *Frontiers in Human Neuroscience, 19*, 1525293. https://doi.org/10.3389/fnhum.2025.1525293

Liu, S., Chen, F., Yin, J., Wang, G., & Yang, L. (2025). Comparative efficacy of robotic exoskeleton and conventional gait training in patients with spinal cord injury: A meta-analysis of randomized controlled trials. *Journal of Neuroengineering and Rehabilitation, 22*(1), 121. https://doi.org/10.1186/s12984-025-01649-1

Maier, M., Rubio Ballester, B., Duff, A., Duarte Oller, E., & Verschure, P. F. M. J. (2019). Effect of specific over nonspecific VR-based rehabilitation on Poststroke motor recovery: A systematic meta-analysis. *Neurorehabilitation and Neural Repair, 33*(2), 112–129. https://doi.org/10.1177/1545968318820169

Mohamad, A., Latir, A., & Manaf, H. (2020). Virtual reality in spinal cord injury rehabilitation: A systematic review of its effectiveness for balance performance and functional mobility. *Healthscope: The Official Research Book of Faculty of Health Sciences, UiTM, 3*(2), 19–24.

Nizamis, K., Athanasiou, A., Almpani, S., Dimitrousis, C., & Astaras, A. (2021). Converging robotic Technologies in Targeted Neural Rehabilitation: A review of emerging solutions and challenges. *Sensors, 21*(6), 2084. https://doi.org/10.3390/s21062084

Ortega-Robles, E., Carino-Escobar, R. I., Cantillo-Negrete, J., & Arias-Carrión, O. (2025). Brain-computer interfaces in Parkinson's disease rehabilitation. *Biomimetics, 10*(8), 488. https://doi.org/10.3390/biomimetics10080488

Peng, Q.-C., Yin, L., & Cao, Y. (2021). Effectiveness of virtual reality in the rehabilitation of motor function of patients with subacute stroke: A meta-analysis. *Frontiers in Neurology, 12*, 639535. https://doi.org/10.3389/fneur.2021.639535

Peng, Y., Wang, J., Liu, Z., Zhong, L., Wen, X., Wang, P., Gong, X., & Liu, H. (2022). The application of brain-computer Interface in upper limb dysfunction after stroke: A systematic review and meta-analysis of randomized controlled trials. *Frontiers in Human Neuroscience, 16*, 798883. https://doi.org/10.3389/fnhum.2022.798883

Phan, H. L., Le, T. H., Lim, J. M., Hwang, C. H., & Koo, K. (2022). Effectiveness of augmented reality in stroke rehabilitation: A meta-analysis. *Applied Sciences, 12*(4), 1848. https://doi.org/10.3390/app12041848

Proietti, T., Ambrosini, E., Pedrocchi, A., & Micera, S. (2022). Wearable robotics for impaired upper-limb assistance and rehabilitation: State of the art and future perspectives. *IEEE Access, PP, 1–1*. https://doi.org/10.1109/ACCESS.2022.3210514

Proietti, T., O'Neill, C., Gerez, L., Cole, T., Mendelowitz, S., Nuckols, K., Hohimer, C., Lin, D., Paganoni, S., & Walsh, C. (2023). Restoring arm function with a soft robotic wearable for individuals with amyotrophic lateral sclerosis. *Science Translational Medicine, 15*(681), eadd1504. https://doi.org/10.1126/scitranslmed.add1504

Pugliese, R., Sala, R., Regondi, S., Beltrami, B., & Lunetta, C. (2022). Emerging technologies for management of patients with amyotrophic lateral sclerosis: From telehealth to assistive robotics and neural interfaces. *Journal of Neurology, 269*(6), 2910–2921. https://doi.org/10.1007/s00415-022-10971-w

Reddy, K. J. (2025). Traditional approaches and limitations. In K. J. Reddy, *innovations in neurocognitive rehabilitation* (pp. 39–51). Springer nature Switzerland. https://doi.org/10.1007/978-3-031-88117-6_3

Rintala, A., Kossi, O., Bonnechère, B., Evers, L., Printemps, E., & Feys, P. (2023). Mobile health applications for improving physical function, physical activity, and quality of life in stroke survivors: A systematic review. *Disability and Rehabilitation, 45*(24), 4001–4015. https://doi.org/10.1080/09638288.2022.2140844

Sarasso, E., Gardoni, A., Tettamanti, A., Agosta, F., Filippi, M., & Corbetta, D. (2022a). Virtual reality balance training to improve balance and mobility in Parkinson's disease: A systematic review and meta-analysis. *Journal of Neurology, 269*(4), 1873–1888. https://doi.org/10.1007/s00415-021-10857-3

Sarasso, E., Gardoni, A., Tettamanti, A., Agosta, F., Filippi, M., & Corbetta, D. (2022b). Virtual reality balance training to improve balance and mobility in Parkinson's disease: A systematic review and meta-analysis. *Journal of Neurology, 269*(4), 1873–1888. https://doi.org/10.1007/s00415-021-10857-3

Shen, Y., Jiang, L., Lai, J., Hu, J., Liang, F., Zhang, X., & Ma, F. (2025). A comprehensive review of rehabilitation approaches for traumatic brain injury: Efficacy and outcomes. *Frontiers in Neurology, 16*, 1608645. https://doi.org/10.3389/fneur.2025.1608645

Szeto, S. G., Wan, H., Alavinia, M., Dukelow, S., & MacNeill, H. (2023). Effect of mobile application types on stroke rehabilitation: A systematic review. *Journal of Neuroengineering and Rehabilitation, 20*(1), 12. https://doi.org/10.1186/s12984-023-01124-9

Tao, Y., Luo, J., Tian, J., Peng, S., Wang, H., Cao, J., Wen, Z., & Zhang, X. (2024). The role of robot-assisted training on rehabilitation outcomes in Parkinson's disease: A systematic review and meta-analysis. *Disability and Rehabilitation, 46*(18), 4049–4067. https://doi.org/10.1080/09638288.2023.2266178

Toh, S. F. M., Fong, K. N., Gonzalez, P. C., & Tang, Y. M. (2023). Application of home-based wearable technologies in physical rehabilitation for stroke: A scoping review. *IEEE Transactions on Neural Systems and Rehabilitation Engineering, 31*, 1614–1623.

Wang, Y., Liang, J., Fang, Y., Yao, D., Zhang, L., Zhou, Y., Wang, Y., Hu, L., Lu, Z., Wang, Y., & Xiao, Z. (2023). Burden of common neurologic diseases in Asian countries, 1990–2019. *Neurology, 100*(21), e2141–e2154. https://doi.org/10.1212/WNL.0000000000207218

Watanabe, H., Marushima, A., Kadone, H., Shimizu, Y., Kubota, S., Hino, T., Sato, M., Ito, Y., Hayakawa, M., Tsurushima, H., Maruo, K., Hada, Y., Ishikawa, E., & Matsumaru, Y. (2021). Efficacy and safety study of wearable cyborg HAL (hybrid assistive limb) in hemiplegic patients with acute stroke (EARLY GAIT study): Protocols for a randomized controlled trial. *Frontiers in Neuroscience, 15*, 666562. https://doi.org/10.3389/fnins.2021.666562

Xie, Y., Yang, Y., Jiang, H., Duan, X.-Y., Gu, L., Qing, W., Zhang, B., & Wang, Y. (2022). Brain-machine interface-based training for improving upper extremity function after stroke: A meta-analysis of randomized controlled trials. *Frontiers in Neuroscience, 16*, 949575.

Yang, F.-A., Lin, C.-L., Huang, W.-C., Wang, H.-Y., Peng, C.-W., & Chen, H.-C. (2023). Effect of robot-assisted gait training on multiple sclerosis: A systematic review and meta-analysis of randomized controlled trials. *Neurorehabilitation and Neural Repair, 37*(4), 228–239.

Yang, J., Gong, Y., Yu, L., Peng, L., Cui, Y., & Huang, H. (2023). *Effect of exoskeleton robot-assisted training on gait function in chronic stroke survivors: A systematic review of randomised controlled trials.* https://doi.org/10.1136/bmjopen-2023-074481

Zhang, H., Jiao, L., Yang, S., Li, H., Jiang, X., Feng, J., Zou, S., Xu, Q., Gu, J., Wang, X., & Wei, B. (2024). Brain-computer interfaces: The innovative key to unlocking neurological conditions. *International Journal of Surgery, 110*(9), 5745–5762. https://doi.org/10.1097/JS9.0000000000002022

Zhang, M., Wang, E., Shan, H., & Zhu, S. (2025). Efficacy evaluation of virtual reality in cognitive and psychological rehabilitation after brain injury: A systematic review and meta-analysis. *Neuropsychiatric Disease and Treatment*, 1455–1468.

Chapter 3
Technological Innovations in Rehabilitation of Musculoskeletal Disabilities

Huma Riaz⑩ and Suman Sheraz⑩

Abstract Musculoskeletal disorders (MSDs) are the dominant source of pain and disability globally. The most prevalent conditions include low back pain, chronic joint pain, arthritis, neck pain, and soft tissue syndromes. Disability from MSDs is staggering, as it has a profound impact on daily functioning, societal and workforce participation, productivity, and healthcare costs worldwide. Recent advances in rehabilitation technology are transforming the assessment and management of MSK conditions. Artificial intelligence–driven systems are increasingly used to support clinical decision-making and personalised rehabilitation planning in musculoskeletal care. Wearable sensors, mobile health applications, and digital therapeutics facilitate continuous monitoring, remote rehabilitation, and patient self-management. Robotic-assisted therapies and virtual or augmented reality further enhance rehabilitation by enabling task-specific training, movement feedback, and engaging in therapeutic environments. This chapter explains how emerging technologies have been extensively integrated in musculoskeletal rehabilitation modalities for precise monitoring, assessment, and intervention. These technologies allow clinicians to quantify movement patterns, track patient progress in real-world environments, and enhance patient engagement in rehabilitation programs.

Keywords Digital therapeutics · Low-Back pain · Musculoskeletal disorders · Musculoskeletal rehabilitation · Osteoarthritis

H. Riaz (✉) · S. Sheraz
Faculty of Rehabilitation and Allied Health Sciences, Riphah International University, Islamabad, Pakistan
e-mail: huma.riaz@riphah.edu.pk

S. Sheraz
e-mail: suman.sheraz@riphah.edu.pk

A. N. Malik et al., *Emerging Technologies in the Rehabilitation of Physical Disabilities*,
SpringerBriefs in Modern Perspectives on Disability Research,
https://doi.org/10.1007/978-981-92-1343-6_3

3.1   Overview of Musculoskeletal Disorders and Disabilities

Musculoskeletal disorders (MSDs) represent a broad group of conditions that affect muscles, bones, joints, ligaments, tendons, and related soft tissues. These disorders are characterised by pain, reduced mobility, impaired neuromuscular performance, and functional limitations that may progress to long-term disability. These musculoskeletal conditions range from acute injuries and inflammatory pathologies to chronic recurrent conditions influenced by mechanical, biological, and psychosocial factors. The disabling impact of MSDs is not limited to structural impairment but also altered motor control, movement patterns, deconditioning, central pain sensitisation, and psychosocial influences. The functional consequences of these MSDs require multidimensional evidence-based rehabilitation approaches in the assessment and management of MSDs.

3.1.1   Classification and Clinical Spectrum

Musculoskeletal disorders (MSDs) comprise a heterogeneous group of almost 150 conditions affecting the locomotor system, including joints, bones, ligaments, muscles, tendons, and associated connective tissue. These conditions affect approximately 1.71 billion people worldwide, leading to transient or permanent limitations in functioning and societal participation (World Health Organization, 2022). MSDs are typically characterised by pain, limited mobility, motor control, dexterity, and general function, thereby reducing work capacity, productivity, and activities of daily living. The clinical spectrum of MSDs expands across the lifespan from childhood to older age, can be of sudden onset as soft tissue injuries, fractures, and others, or of gradual or long-term onset as chronic low back pain and osteoarthritis associated with pain and activity limitations. The classification of MSDs comprises of conditions related to joints (osteoarthritis, rheumatoid arthritis, gout, spondyloarthritis), bones (traumatic fractures, osteopenia, osteoporosis and associated fragility fractures), muscles (sarcopenia), and multiple body systems or areas, such as regional (e.g. back and neck pain) and widespread (e.g. fibromyalgia) pain conditions, inflammatory diseases such as connective tissue diseases (Kiesel et al., 2024).

3.2   Global Burden of MSDs and Related Disability

In 2020, MSDs were ranked as the second highest cause of non-fatal disability, listed as the leading global contributors of physical disability with significant limitations in mobility and dexterity. MSDs progress to lower levels of well-being, early work retirement, and compromised societal participation. (James et al., 2018). Evidence from the burden estimates, epidemiological studies, and clinical research explains

MSDs as the leading determinant of disability, reporting disability outcomes as prevalence, Years Lived with Disability (YLD), Disability-Adjusted Life Years (DALYs), functional impairments, correlations or impact on quality of life.

The number of people living with MSDs is rapidly increasing due to global ageing and population growth, contributing to the increased burden of disability. The Global Burden of Disease (GBD) data analysis showed that worldwide 1.71 billion people are living with MSDs, including low back pain, neck pain, osteoarthritis, rheumatoid arthritis, amputation and other injuries (James et al., 2018). MSDs are affecting a large number of people in all parts of the world, with 441 million in high-income countries, followed by 427 million in the WHO Western Pacific Region and 369 million in the South-East Asia Region (Cieza et al., 2021). MSDs are the second biggest cause of YLDs globally, after mental health disease, making approximately 179 million YLDs, accounting for almost 17% of all YLDs and 150.08 million DALYs of MSDs. Particularly, in adolescence and young adults, the MSDs have emerged as the third leading cause of DALYs in the last three decades, with a high proportion of DALYs from high BMI than from occupational ergonomics and smoking (Guan et al., 2023; Liu et al., 2022). Similarly, over the past three decades, population ageing has become the leading contributor to increases in incidence, prevalence, and DALYs. Males were more affected by population ageing, specifically in high-and high-middle sociodemographic index countries, whereas females were more affected in low-to middle-sociodemographic index countries (Guan et al., 2025).

The GBD study (2019) identifies occupational risk, tobacco use, and high body mass index (BMI) as the highest contributors to MSDs, with high BMI being liable for 7.3 million YLDs across 192 countries and territories. MSDs have a significantly high socioeconomic impact with both direct and indirect costs on patients and healthcare systems. Estimates of global health and economic outcomes attributable to high BMI showed a total cost of $180.7 billion, including $120.2 billion in productivity losses and $60.5 billion in healthcare costs (Chen et al., 2023). Further, the prevalence of MSDs is found to be greater in women, advanced age, and higher income countries, with a positive correlation with national income level and medical density (Bouziri et al., 2023).

Around the world, musculoskeletal conditions, such as low back pain, osteoarthritis, and neck pain, are identified as the leading global causes of disability measured as YLDs. In contrast, in 2020, the other MSDs were ranked the sixth leading cause of YLDs (Gill et al., 2023). Low back pain (LBP) is the major public health concern globally, and the leading cause of functional limitation, disability, and work absenteeism across all age groups and socioeconomic strata, with high clinical and economic burden to individuals, health systems, and society, in low-, middle-, and high-income countries (Fatoye et al., 2023). Evidence states that in 39 countries, the median length of hospital stay due to LBP is 6.2 days for 'dorsalgia' and 5.4 days for 'intervertebral disc disorders' (Melman et al., 2023). The findings from GDB 2019 indicate that the global prevalence of LBP is higher among females than among males, and the number of prevalent cases increases with age, peaking at 45–54 years for both sexes (Chen et al., 2022).

Another major public health problem is neck pain, which is multifactorial with significant biological and psychological risk factors. In 2019, the age-standardised global prevalence rate of neck pain was 27/1000 population, with a remarkable economic burden including medical and rehabilitation costs, reduced work productivity, and job-related problems (Kazeminasab et al., 2022a). Neck pain is the most common MSD among adolescents and young adults, globally accounting for 15.4% of DALYs. The GBD 2019 data indicate a decreasing trend in age-standardised prevalence, incidence, and DALYs for neck pain over the next 30 years.

Another highly prevalent condition is Osteoarthritis (OA), which is a chronic, progressive joint disorder with irreversible structural changes in the hands, knees, hips, and feet. It is estimated to have 30.3 million prevalent cases and 1.5 million incident cases reported in 2017, and it carries a tremendous health and economic burden (Cao et al., 2024). These descriptive estimates underscore its position as the leading cause of disability worldwide. This MSD progresses to activity limitations, formidable personal suffering, deformity, and disability in the ageing population. The burden of knee OA has increased globally from 1991 to 2021, with a faster rate among females than among males. The KOA symptoms are increasing more rapidly in the 35–39-year age group, whereas the prevalence and DALYs are increasing more rapidly in the 45–49-year age group (B.-F. Zhang et al., 2025). Overall, the rising prevalence and long-term consequences of musculoskeletal disorders continue to pose significant challenges for healthcare systems globally.

3.3 Musculoskeletal Disability within the International Classification of Functioning (ICF) Framework

As per the International Classification of Functioning Framework proposed by WHO, disability is not merely a medical condition but a complex and negative interplay between a person's health condition and contextual factors (including personal and environmental). Disability is a multidimensional phenomenon encompassing impairments, activity limitations, and societal restrictions. MSDs are the health conditions that simultaneously affect all domains of disability.

Within the International Classification of Functioning, Disability, and Health (ICF) framework proposed by WHO, the musculoskeletal disability progressing from MSDs is defined through a multidimensional, biopsychosocial view of health that goes beyond structural pathology. In this framework, disability is conceptualised as the dynamic interaction between health condition and its specific contextual factors. The disability is systematically categorised into several interacting components, including effects of health condition on body function and structures, activities, social participation, and contextual factors. In MSD-related physical disabilities, the impairment-level evaluation alone cannot predict the degree of functional limitation or participation restriction, and the ICF framework highlights and guides comprehensive assessment by addressing activity limitations, participation restrictions, and contextual factors. For example, individuals with Knee OA, having similar severity of structural degeneration on radiographs, may demonstrate variation in pain-related disability and mobility limitations, underscoring the insufficient power of solely structural impairments in predicting the degree of disability. The ICF can also specify the domains of disability related to individuals with MSDs by using ICF Core Sets (Escorpizo, 2014a; Hill et al., 2025).

3.4 Impairments and Functional Limitations of Common MSDs

Common MSDs such as low back pain and neck pain are major contributors to disability worldwide. They present with a range of physical impairments that contribute to activity limitations and participation restrictions. LBP is a common MSD, encompassing three sources of pain, (1) the axial lumbosacral pain; refers to pain in the lumbar region (L1–5) and sacral spine (S1) to sacrococcygeal junction, (2) radicular pain, in dermatomal distribution due to nerve root or dorsal root ganglion irritation, travelling down to lower extremity, and (3) the referred pain, following non-dermatomal trajectory, in the region remote from the source of pain. LBP can also be classified by clinical course or chronicity as acute (<6 weeks), subacute (6–12 weeks), or chronic (>12 weeks). The causes of LBP are varied, including myofascial, facet joint-mediated, discogenic pain, and others. Non-specific chronic LBP, where no pathoanatomic cause can be identified, is the most common type of LBP and a major public, economic, occupational, and social burden (Urits et al., 2019).

Chronic LBP presents with impairments that extend beyond pain intensity, including biomechanical, neuromuscular, postural, mobility, and cognitive domains. A significant reduction in lumbar spine movements, alteration in paraspinal muscles morphology, either weakness, atrophy, or tightness, impaired postural stability with greater postural sway and poor balance control exacerbated by higher pain severity, altered gait patterns, including reduced step length, slower gait speed, and compromised trunk-pelvis coordination (Errabity et al., 2023). These multidimensional impairments translate into functional limitations (sit-to-stand transition, lifting tasks, prolonged standing, and work-related activities), participation restrictions (attending work, limitations in family and domestic roles, restricted leisure activities, sports participation), and work-related activities. Such activity and social limitation, if compounded by psychosocial factors such as fear avoidance beliefs, can intensify movement avoidance at the lumbar spine with more deconditioning, further contributing to physical disability (Kalli et al., 2025; Sun et al., 2023).

Similarly, neck pain is among the top five chronic pain conditions, in terms of prevalence and YLDs, with considerable personal and socioeconomic burden. It commonly affects two-thirds or more of the general population at one point in their life. Cervical spine pain or neck pain is the pain perceived in the posterior part of the neck from the superior nuchal line to the first thoracic spinous process (Misailidou et al., 2010). It can be classified as acute (<1 month) or chronic (>6 months). Conservative or pharmacological treatments resolve most cases of acute neck pain, whereas in nearly 50%, it persists with some degree of pain or frequent recurrence. There are certain risks associated with neck pain. Evidence has shown that individual/biological factors (ageing, genetic, pre-existing neuromuscular or autoimmune disorders) and psychological factors (stress, anxiety, depression, some cognitive factors, and sleep problems) both contribute to the development of neck pain (Kazeminasab et al., 2022b).

There is a moderate correlation between neck pain intensity and disability. Chronic neck pain, including chronic non-specific neck pain or mechanical neck pain, is associated with multidimensional impairments, including pain, restricted cervical spine mobility, hypo-or hyperactive neck muscles with trigger points, poor sensorimotor control, impaired postural control and balance, and psychosocial, cognitive, and affective impairments. The constellation of these structural and functional impairments leads to reduced ADLs, functional performance, work capacity, and productivity and compromised quality of life (Fejer & Hartvigsen, 2008; Tanik & Ozer Kaya, 2024).

Osteoarthritis is another degenerative condition associated with a range of structural and functional impairments. These impairments progressively affect mobility and daily functioning. OA is a complex, multifactorial joint pathology caused by inflammatory and metabolic processes that contribute to progressive joint damage. It is characterised by degeneration of articular cartilage and persistent joint pain, which ultimately leads to functional limitations and disability. Individuals with knee OA frequently experience persistent joint pain, stiffness, reduced range of motion, and decreased quadriceps muscle strength. These collectively compromise patients' knee stability and movement efficiency. Altered gait mechanics, reduced shock absorption, and impaired balance are also commonly observed, contributing to limitations in activities such as walking, stair climbing, and rising from a seated position (Langworthy et al., 2024; Uivaraseanu et al., 2022).

These conditions often lead to persistent pain, reduced spinal mobility, muscle weakness, and impaired movement control, which limit the ability to perform daily activities and functional tasks. Over time, recurrent symptoms and functional limitations may contribute to reduced physical activity, work absenteeism, and decreased quality of life, thereby increasing the overall burden of disability. Persistent pain and movement limitations may also lead to compensatory movement patterns and reduced physical conditioning, further aggravating symptoms. Fear of movement, reduced confidence in physical activity, and prolonged inactivity lead to a transition from acute symptoms to chronic disability. As a result, these conditions not only affect physical functioning but also influence psychological well-being, social participation, and overall health outcomes (Shimizu et al., 2022).

3.5 Mechanisms of Recovery in Musculoskeletal Rehabilitation

Rehabilitation has been defined in the World Health Report 2011 as "a set of measures that assist individuals who experience, or are likely to experience, disability to achieve and maintain optimal functioning in interaction with their environment" (World Health Organization, 2011). MSD-related disabilities are amenable to rehabilitation. It is a multifactorial process aimed at optimising the individual's functioning. There is a plethora of evidence suggesting the effectiveness of various rehabilitation approaches for the diagnosis and management of disabilities caused by chronic musculoskeletal conditions. The rehabilitation approaches consider all the disease and non-disease-related factors. For example, for rehabilitation of disabilities related to arthritis, the factors such as socioeconomic status, medication intake, depression, and self-efficacy are considered in management strategies (Escorpizo, 2014b).

3.5.1 Biological Mechanisms

In musculoskeletal tissues, the repair after acute or chronic injuries is a significant clinical challenge. The recovery mechanism involves cellular repair, inflammation resolution, and mechanotransduction. A very complex, interlinked sequence of cellular, humoral, and often vascular events explains the healing of musculoskeletal tissues, restoring their structure and function after injury. In vascularised tissues such as bone, muscle, and dense fibrous tissue, the healing process follows a continuous sequence of events, including inflammation, repair, and remodelling (Buckwalter, 1996). In this context, recent evidence explains the molecular regulators involved in the regeneration of musculoskeletal tissues. The molecular cues, such as biochemical and biophysical signals, orchestrate cellular repair and regeneration. Molecular cues, such as growth factors, cytokines, and non-coding RNAs, particularly microRNAs, as well as critical pathways, such as TGF-β, FGF, and NF-κB, regulate healing and can promote tissue regeneration (Peniche Silva et al., 2026). These biochemical pathways of pain and tissue maintenance can be influenced by targeted rehabilitation. Additionally, Movement and loading responses in musculoskeletal tissue through therapeutic exercises stimulate mechanosensitive pathways and can greatly influence musculoskeletal tissue healing.

3.5.2 Neuromuscular Mechanisms

In MSK rehabilitation, neural components are important in the early phase of strength recovery. Neural adaptation includes motor unit activation, increased neural firing, and coordinated synchronisations, which support motor control in the early phase of rehabilitation. Muscle hypertrophy enhances muscle strength and power, providing joint stability and mobility during recovery. The gradual and progressive eccentric loading of the neuromuscular system improves soft tissue stiffness and increases load tolerance to prevent re-injury (Cakmak, 2023). In MSK injury, individual motor control constraints include pain, limited movement, and weakness, which impose on the nervous system the need to develop compensatory strategies. It is important to focus on restoring normal movement, synergistic patterns, and functional activities. In the early phase of recovery from MSK injury, feedback plays a significant role in accelerating recovery. Feedback includes the intrinsic, which refers to vision and proprioception, and the extrinsic, which refers to verbal, visual, and biofeedback, to optimise functional restoration.

3.5.3 Biomechanical Optimisation

Biomedical optimisation is the mechanism by which the body applies appropriate force distributions to achieve smooth, effective movements across multiple joints and muscles. Injury in MSK leads to abnormal movement patterns and compensatory strategies due to altered force distribution, altered kinematics, and muscle incoordination. In MSK rehabilitation, training should be targeted to reduce joint stress, restore alignment, enhance stability and coordination, improve mechanical efficiency, equal load sharing, and reduce the risk of recurrent injury. Progressive loading is a key factor in MSK rehabilitation; under-and overloading can delay recovery. The rehabilitation should focus on a gradual shift in load to promote resilience and remodelling.

3.5.4 Psychosocial Modulation

Psychological modulation in MSK injury refers to the impact of emotional and cognitive factors on movement patterns and behaviours. Recovery is not only linked to the MSK system; the brain continuously processes signals from the body to map post-injury. Fear of pain, catastrophising, recurrence of injury, kinesiophobia, lack of confidence, and history of previous events lead to alteration in motor control. Motor control has been manipulated through attention, motivation, cognitive load, and fear. The environment should be provided to enable activities to be performed safely and without fear.

In MSK recovery, the biopsychosocial model should be followed to provide comprehensive rehabilitation to address cognitive, emotional, and physical aspects simultaneously for better and speedy recovery. In case of non-compliance, these maladaptive beliefs lead to deconditioning and long-term disability. Rehabilitation should integrate patient education, gradual exposure to movement, reassurance and confidence-building, and cognitive-behavioural strategies to minimise fear and phobias and enhance confidence. Cognitive and emotional aspects significantly enhance recovery, functionality, and mobility in long-term rehabilitation (Cakmak, 2023).

3.6 Technological Innovations in Musculoskeletal Assessment

Technological advancements have drastically influenced and revolutionised musculoskeletal assessment and monitoring. Assessment forms the framework for musculoskeletal rehabilitation by guiding diagnosis, goal setting, and prognosis of the MSK condition. Clinically, the rehabilitation plan is designed for patients based on outcomes such as pain, joint mobility, muscle performance, and functional capacity.

Conventional musculoskeletal assessment relies on patient-reported outcomes, clinical evaluation, and manual measurement tools. These assessment approaches are clinically valuable but are somewhat affected by the subjective interpretation of the outcome assessor. There is a risk of compromised inter-rater reliability, particularly with outcomes such as range of motion, manual muscle testing, and postural analysis. Moreover, most evaluations are conducted within controlled clinical environments and do not reflect real-world functional demands. Conventional clinical tools do not adequately capture certain biomechanical deviations, compensatory movement patterns, and variations in patients' symptom patterns. Also, they fail to capture longitudinal trends and patients' adherence to the intervention protocol. The objectivity, precision, and ecological validity of assessment systems have improved with the integration of technology. Also, these systems help with quantitative movement analysis, continuous patient monitoring, and adherence, thereby improving the assessment of MSK impairments and functional limitations.

Current evidence, though inconsistent, indicates that digital physiotherapy assessments can be comparable to traditional face-to-face assessments. For many outcome measures, both assessment methods have yielded similar validity and reliability results. Higher levels of validity have been reported for assessments such as extremity Range of Motion (ROM), Patient-Reported Outcome Measures (PROMs), and aspects of clinical management. In contrast, lower validity has been observed in areas such as posture assessment and certain clinical observations (Bernhardsson et al., 2023). MSK assessment typically focuses on core outcome measure domains such as pain, mobility, functional ability, and quality of life. Multiple emerging technologies, ranging from digital software and wearable devices to AI-driven tools, can augment each domain. Emerging technologies provide objective measurements, continuous monitoring, and personalised feedback, enhancing the accuracy and effectiveness of MSK assessment. The role of technological innovations in assessing and monitoring MSK outcomes across different domains is explained in the following section.

3.6.1 Pain and Related Factors

Wearable sensors enable real-time, longitudinal measurement of pain-associated physiological changes via devices such as smartwatches and wristbands. Currently, there is no objective gold-standard tool for accurately measuring pain. However, wearable sensors can track correlated physiological signals to assess pain in paediatric and adult populations. These physiological parameters include heart rate, heart rate variability, electrocardiography, electromyography, photoplethysmography, skin temperature, and autonomic, muscular, and neural system activity measured via electroencephalography. This is especially important for measuring pain when self-reporting is limited or impossible, e.g. in comatose patients, infants, and cognitively impaired patients. Combining signals from multiple sensors improves pain assessment compared with a single-sensor approach (Semiz et al., 2026).

Pain assessment is also conducted through AI-based interventions that support pain recognition and prediction. AI-based models also help assess pain intensity by identifying facial features. Different AI-based techniques are currently being used to assess pain (M. Zhang et al., 2023a). Machine Learning (ML) recognises pain signals, classifies them by severity, and identifies pain patterns from clinical data. Multiple ML models have been developed that analyse facial expressions for pain recognition. Data Mining techniques are applied to large datasets to uncover patterns and relationships in pain responses that may support assessment. Natural Language Processing (NLP) is used to extract and analyse pain-related information from patients' reports and clinical notes. It reads and analyses patients' pain-related data. It identifies pain-related patterns, experiences, and descriptions of pain by their intensity, location, and duration. Another technique is computer vision, which analyses visual data such as digital images and videos of patients. It analyses signals such as facial expressions, body movements, and movement patterns (Huo et al., 2024).

The Internet of Things (IoT) allows continuous monitoring of physiological and behavioural signals. Devices with IoT technology can track patients' vital signs and movement patterns. These devices transmit data to cloud platforms, allowing clinicians to monitor patients remotely and identify pain trends over time. Integrating these with AI can detect pattern changes, support timely interventions, and complement traditional pain scales, especially for chronic patients or those with communication difficulties (El-Tallawy et al., 2024).

Ecological Momentary Assessment (EMA) is also a data collection method that collects real-time data using digital devices such as smartphones, tablets, or mobile applications. This assessment method can capture environmental factors, activity levels, mood, and pain-related triggers. EMA also helps collect data at multiple points, thereby recording trends over time. It can also be integrated with wearable and IoT devices for comprehensive data recording.

There are multiple advantages of these technology-assisted tools over conventional pain assessment methods. It allows for a more objective assessment of pain, especially in patients who are non-verbal or have impaired communication. It helps detect pain early, enabling timely intervention. It also differentiates genuine pain from simulated expressions. However, with these technology-based assessment methods, there is always a risk of error and inaccuracy that can lead to incorrect diagnoses and management decisions (El-Tallawy et al., 2024).

3.6.2 Range of Motion (ROM) and Joint Mobility

Advances in technology have also enabled accurate measurement of the ROM with digital and AI-driven goniometers. These devices facilitate remote assessment and telerehabilitation, helping the patients with self-assessment. Camera-integrated AI-driven ROM assessment tools detect landmarks and calculate joint angles, providing a more accurate ROM assessment. DETROM is an AI-driven goniometer that assesses ROM, specifically knee flexion and extension. A meta-analysis comparing traditional goniometry with digital ROM devices demonstrated minimal measurement bias ($\approx$ $-0.25°$) between the two methods (Koul et al., 2025). Reliability of digital tools is also high, with ICCs ranging from 0.62 to 0.99, comparable to standard goniometric assessment (Shepherd et al., 2024). Measurement consistency was maintained across different planes of movement (flexion, abduction, rotation), supporting clinical applicability of digital tools. Another meta-analysis highlighted a variety of technologies for assessing shoulder ROM. Inertial Measurement Units (IMUs) and electrogoniometers are wearable solutions that incorporate accelerometers and gyroscopes that provide real-time ROM data and are suitable for both upper-and lower-limb assessment. Smartphone-based applications utilising built-in motion sensors or camera-based analysis, and virtual reality–integrated motion-tracking systems are other technologies used for the same evaluation of shoulder joint ROM (P. Chen et al., 2025).

Computer vision (CV) solutions utilise 2D and 3D camera technologies, RGB-Depth sensors, and smartphone cameras to track body movements and extract joint angles using image processing and machine learning algorithms. By automatically detecting landmarks and calculating ROM, CV-based assessments provide objective, quantitative measurements that can be used in both clinical and remote rehabilitation settings (Aleksić, 2023). These technologies allow continuous tracking of ROM and functional outcomes, supporting personalised rehabilitation programs and enhancing patient adherence.

3.6.3 Muscle Strength and Neuromuscular Function

Muscle strength and neuromuscular function are central determinants of joint stability and functional independence. Emerging technologies help quantify force production objectively, identify muscle-activation patterns, and assess neuromotor control. Digital handheld dynamometers are used to assess muscle force production, as traditionally assessed with manual muscle testing. The Gripwise and ActivForce digital dynamometers are examples of devices used to monitor patients' muscle strength over time (Benatti de Oliveira et al., 2025).

EMG monitoring systems detect motor-unit recruitment patterns, muscle-activation amplitude, fatigue responses, and neuromuscular coordination. Wireless and wearable EMG sensors enable assessment during functional tasks, directly linking neural activation to movement performance. EMG also estimates muscle power, movement velocity, and functional strength when combined with wearable sensors (IMUs). Advanced EMG applications include force-estimation algorithms that can indirectly calculate muscle-generated torque during movement. Symmetry and limb deficit analysis help assess muscular imbalances or unilateral weakness. Additionally, fatigue detection models monitor changes in muscle-activation patterns over time (Mokri et al., 2022; Moran et al., 2022).

3.6.4 *Movement Quality and Biomechanical Analysis*

Accurate assessment of movement quality is essential for understanding functional performance and detecting compensatory strategies. 3D Motion Capture Systems are considered the gold standard in gait analysis. They use multiple cameras and reflective markers (or markerless depth sensors) to track joint positions and body segments in three dimensions. Smart Insoles and Pressure Sensors measure load distribution, centre of pressure, and plantar pressure patterns during standing and locomotion. This technology helps detect abnormal loading, assess balance, and monitor rehabilitation progress in postsurgical patients and those with MSDs (Santos et al., 2024; Scataglini et al., 2024; Wang et al., 2025).

Vision-based sensor systems use cameras and depth sensors to record human gait without markers or wearable devices. Machine learning models help provide step length, gait speed, and other parameters from video data. The accuracy and reliability of these systems are improved through proper camera placement and strong algorithms. These systems support real-time gait analysis for both clinical and research applications, enabling comprehensive assessment of movement quality and functional performance (Han et al., 2025).

AI-powered posture assessment systems use computer vision and deep learning to analyse human posture. They analyse posture from standard images or video, extracting anatomical key points and body alignment patterns. One such example is the ZarifTool system, which can generate personalised physiotherapy recommendations based on posture deviations through customised corrective exercises (Al Dibs et al., 2025). Multibody dynamics-based musculoskeletal modelling helps quantify internal joint forces, muscle activations, and joint moments during gait by integrating motion capture, force, and EMG data. Combined with smart insoles, AI posture analysis, and gait systems, it supports objective assessment of movement quality and functional performance (Abdullah et al., 2024).

3.6.5 Functional Performance Measures

Continuous, objective assessment of functional performance measures is now also being done using digital technologies. Tools such as digital 6-min walk tests (6-MWT) and sit-to-stand trackers assess endurance, mobility, and lower-limb strength. 6-MWT performance is monitored using motion and inertial sensors embedded in various devices (Pires et al., 2022). Patients can record videos of sit-to-stand tests, which are analysed using a computer vision algorithm to detect movement parameters. Wearable activity monitors record daily step count, intensity, and movement patterns. The collected data can be compiled into a dashboard that helps therapists track progress and update the intervention plan without in-person visits to patients. Frailty metres are also digital measures of functional performance that use wearable sensors integrated in wrist-worn bands. These devices assess parameters such as speed, range of motion, power, and exhaustion as markers of functional performance, helping with the levels of frailty. Additionally, return-to-activity readiness models use sensor and performance data to predict patient readiness for higher-level functional tasks, sports participation, or work activities (Hwang et al., 2024).

3.6.6 Patient-Reported and Psychosocial Outcomes

Monitoring functional and psychosocial outcome measures has greatly improved with the digital tool. App-based disability scales allow individuals to self-report pain, stiffness, and daily activity limitations. This helps clinicians track symptom fluctuations that may not be captured during clinic visits (Kang et al., 2023). Tracking fear avoidance and kinesiophobia can identify patients who are reluctant to get engaged in rehabilitation due to anxiety and movement-related fear. Wearable devices and smartphone apps also track sleep and activity behaviour. Also, patient adherence data provide feedback on exercise completion and engagement with the intervention protocol, helping clinicians with in-depth monitoring and tailoring the intervention to patients' needs (Feng et al., 2026).

3.7 Technological Innovations in Musculoskeletal Management

Technological advances have played a major role in transforming how we support and rehabilitate people living with musculoskeletal disorders and related disabilities. Advances in technology have played a major role in the rehabilitation of individuals living with MSDs and other disabilities. Traditional approaches were more focused on physical symptoms, such as pain management and movement restoration. Conventional rehabilitation approaches had some limitations, including subjective assessment, episodic care, generic and setting-bound interventions, and accessibility barriers.

Whereas the contemporary rehabilitation mechanisms address both the physical and psychosocial aspects of recovery by combining physical therapy, regenerative medicine, innovative technologies, and psychological support, this multidisciplinary approach to integrating cutting-edge interventions has enhanced treatment precision, improved recovery outcomes, enabled continuous monitoring and remote access, and enabled personalised, tailored rehabilitation plans tailored to specific patient needs. The synergy between the latest research, clinical expertise, and emerging therapeutic modalities and technologies holds immense promise for improved functional restoration, enhanced mobility, reduced disability, and better quality of life for patients with MSDs.

3.7.1 Artificial Intelligence-Driven Systems

AI algorithms estimate body posture and biomechanics and objectively quantify movements using smartphone cameras and low-cost sensors, enabling markerless motion capture. Machine learning (ML) applications can transform clinician-dependent assessment into automated, data-driven insights. The evidence synthesis from a recent systematic review applying ML approaches to the rehabilitation of adults with non-specific LBP can facilitate health professionals' clinical decision-making and guide patients in self-managing LBP. ML application in this study was classified in three domains: (1) ML models to classify movement patterns into normal versus abnormal based on input from motion capture or wearable sensor, (2) ML models predicting risk of poor rehabilitation outcomes and prolonged disability, (3) ML models using supervised learning to provide corrections and adjust exercise difficulty. This review also guides the protocol design of an ML-assisted movement retraining application, using real-time feedback to guide postural control or gait as per FITT guidelines; interactive, continuously sensor-monitored sessions 3–5 days/ week, moderate intensity, 20–40 min for 6–12 weeks (Amorim et al., 2021).

Another systematic review (SR) and meta-analysis (MA) finding indicates that AI-assisted physiotherapy, through enhanced personalisation and adherence, can provide modest short-term benefits in reducing pain and disability in non-specific LBP. This review included studies primarily involving AI-guided exercise prescription systems, ML-based progression algorithms, motion-tracking platforms with automated feedback, and app-based rehab platforms. The suggested dosage was reported as: 2–5 sessions/week with continuous remote monitoring, moderate intensity exercises with progression determined by pain and range of motion scores, 20–45 min for 6–12 weeks (Kapil et al., 2025). Systematic evidence shows that AI technologies, ranging from supervised ML to computer vision, wearable sensor analytics, or mobile platforms, enhance all domains of physical rehabilitation, including assessment, intervention delivery, performance optimisation, and outcome prediction. Particularly in interventions, algorithms can provide adaptive exercise prescriptions, progression criteria, and real-time feedback to correct movement in patients with MSD pain (Sumner et al., 2023).

Another rigorous scoping review explains that AI approaches, including ML, DL, and NLP, align with precision health paradigms and are promising for facilitating pain management by providing objective data, adaptive self-management instructions, and predictive insights. The study findings show that AI can provide objective digital biomarkers beyond subjective measures; for instance, in LBP, automated pain detectors can support clinical pain rating scales such as the Numeric Pain Rating Scale or the Visual Analogue Scale. For neck pain, AI can be trained on sensor data that captures movement irregularities or postural instability. Similarly, for OA, AI-enabled apps can track daily pain fluctuations and offer activity adaptation (M. Zhang et al., 2023b).

AI technologies in knee OA care can be used for ongoing monitoring, risk prediction, adherence support, and workflow optimisation. Data on pain levels, joint mobility metrics, and physical activity patterns can be collected by AI-based remote monitoring systems using wearable sensors and mobile platforms. Further, patients with knee OA can be educated through self-management skills, guided home exercise instruction, and reminders about medication delivery and lifestyle by AI-driven chatbots and virtual health assistants. These tools enhance self-efficacy in chronic pain conditions and increase adherence to prescribed rehabilitation programs (M. Zhang et al., 2023b).

3.7.2 Mobile Health-Based Applications

Mobile health (mHealth)-based rehabilitation programs delivered through smartphones and connected devices have emerged as patient-centred digital interventions for patient education, exercise delivery, symptom monitoring, and behavioural support. mHealth-based interventions facilitate musculoskeletal recovery through all mechanisms. The neuromuscular domain is targeted through strength and endurance training, guided motor control activities, and progressive functional mobility retraining. Psychosocial modulation is facilitated by education, behaviour reinforcement, self-efficacy, and reduced fear avoidance. Biomechanical recovery is done with movement pattern training and postural correction. Lastly, the auto prompts, progress visualisation, and habit formation are achieved through behavioural adherence mechanisms.

An SR and MA of RCTs were conducted to evaluate the effects of mHealth interventions on the management of chronic LBP in low- and middle-income countries (LMICs). The findings indicate that mHealth interventions comprising an app-delivered exercise program, pain education modules, and self-monitoring reinforcement showed a modest reduction in pain intensity compared with control/usual care and little to moderate improvement in disability outcomes. Behavioural education with exercise showed stronger effects than the exercise-only format (Rani et al.,

2025). Contrary to this, another SR and MA have shared the inconclusive findings that mHealth interventions alone, without therapist direct supervision, are not effective for pain and disability associated with chronic LBP. The included studies have used app-based delivery of education, self-management instructions, exercise programs, symptom tracking, and behavioural support parts, and the intervention periods ranged from 6 weeks to 12 months (de Melo Santana et al., 2023).

mHealth interventions are also being used in the management of chronic neck pain, causing global disability. A recent SR & MA has synthesised findings from 10 RCTs with 825 participants, and the analysis indicated that mHealth interventions were effective for neck pain and disability only when compared with usual or no care, not other physical therapies. The heterogeneity of intervention design, small sample size, and lack of validity in some app features are the limitations (Xie et al., 2025). Another recent SR & MA, including 6 RCTs and 381 participants with chronic neck pain, evaluated the comparative effectiveness of mHealth interventions and conventional exercise interventions (He et al., 2025a). The analysis demonstrated that mHealth-supported exercise was equally effective as supervised traditional exercise in reducing pain and improving function, while showing superior outcomes compared with unsupervised home exercise programs. The findings from these reviews show that mHealth tools can make rehabilitation more accessible, improve exercise adherence, and produce results similar to those of a regular exercise program (He et al., 2025b).

The effectiveness of mHealth interventions for patients with hip or knee osteoarthritis was reviewed, comparing digital interventions with usual or conventional care. A small to moderate improvement was found in both pain intensity and functional disability. The benefits were greater when mHealth platforms provided interactive, feedback-enabled designs through structured guidance rather than passive information alone. Despite some methodological limitations, the review provided moderate-quality evidence supporting mHealth as an effective adjunct in conservative OA management, particularly for improving patient-reported pain and function (Mapinduzi et al., 2025). Similarly, another review reports modest advantages of mHealth-based exercise interventions over traditional exercise in reducing pain and functional disability and improving quality of life in patients with knee OA. The non-inferior effects were observed when mHealth interventions were compared with traditional supervised interventions, whereas superior effects were reported when compared with unsupervised conventional exercises (Tang et al., 2025).

3.7.3 Digital Therapeutics (DTx)

Digital therapeutics (DTx) are evidence-based software interventions regulated as medical devices and most often used in home care to manage musculoskeletal conditions. The enabling core technologies used in DTx are AI-driven motion analysis for objective functional assessments, wearable sensors (smart insoles, EMG, IMUs) for continuous biomechanical testing about joint angles, muscle activity patterns, and step symmetry, cloud-based ecosystems supporting secure data integration, analytics, and feedback loops, telerehabilitation platforms for remote care, and AR/VR environments for interactive exercise guidance. A recent narrative review on DTx in musculoskeletal care has synthesised evidence from over 75 studies, including clinical trials and validation studies. The findings suggest that DTx has enhanced patient adherence to rehabilitation programs by 15–40%, reduced healthcare costs by 30–40% and has maintained or increased functional outcomes (including Pain and ROM) across the orthopaedic procedures (Lee et al., 2025).

3.7.4 Robotic-Assisted Therapies

The use of robotics in musculoskeletal rehabilitation is supported by robust evidence of effects on functional improvement and pain relief, particularly in OA. Robotic devices are being used and evaluated, particularly for integration with AI and intelligent feedback systems. These devices are effective modalities for personalised MSD rehabilitation rather than routine care. Robotic interventions in musculoskeletal rehabilitation follow some specific mechanisms. Through biomechanical augmentation, robots can achieve precise joint mobilisation and movement control. Neuromuscular retraining using balance platforms: repetitive task-specific training with feedback for lower extremity arthritis and other MSDs. Nociceptive modulation can be achieved with robotic massage for chronic LBP, producing mechanical stimulation and reducing pain perception.

Some robotic interventions, including exoskeletons and assistive devices, are suitable for the rehabilitation of physical disabilities, as they enhance motor function, strength, coordination, and dexterity compromised in MSDs (Banyai & Brişan, 2024). A high-level synthesis supporting robotic interventions for musculoskeletal rehabilitation has found that Single-Joint Rehab Robots are better for ROM improvement than Robotic exoskeletons, which are ranked highly for pain relief (Luo et al., 2025). The application of robotic technologies has been less studied in the context of chronic LBP. An RCT using a therapeutic massage robot, an automated mechanised soft-tissue device (ADAMO), in patients with non-specific chronic LBP showed significant improvement in pain, measured by VAS, and disability, assessed by the Oswestry Disability Index. There are some emerging personalised soft wearable robotic backpacks under testing, to explore their effects on pain and daily activities in patients with chronic LBP (Marín-Méndez et al., 2021).

OA has the strongest robotic evidence, particularly post arthroplasty. An RCT was conducted with older adults undergoing arthroplasty who received robot-assisted knee rehabilitation training, and significant improvements in knee ROM, functional ability, recovery, and ADLs were observed. Patients with hip or knee joint arthritis also experience balance deficits and an increased risk of falls (Castelli et al., 2023a). In this context, evidence from RCT findings demonstrates robotic benefits for mobility impairments in OA-related disability. The technology used for intervention, the Hunova® robotic balance platform, has shown significant improvements in motor performance, balance, gait parameters, and reduced fatigue in OA patients post-knee replacement (Castelli et al., 2023b).

Robotics are rarely used in isolation; it is mostly integrated with conventional physical therapy. The dosage of robotic training, consistently informed by evidence, consists of higher-frequency, feedback-guided sessions in postoperative conditions, while in chronic pain conditions, short-term training (4–6 weeks) of 20–45 min.

3.7.5   Virtual (VR) and Augmented Reality (AR) Intervention

VR and AR improve musculoskeletal rehabilitation through three main outcomes. First, the virtual environment reduces short-term pain intensity through desensitisation of pain receptors, active engagement during training, reduced fear-avoidant behaviours, and cognitive pain modulation. The virtual environment prepares patients for graded exposure to painful movement and improves adherence with exercise therapy. Secondly, functional outcomes are achieved in VR-supported training through improved range of motion, enhanced postural control, better dynamic motor control, and increased functional performance. Thirdly, VR has a positive impact on psychological impairments, including kinesiophobia, pain catastrophising, training adherence, and fun and motivation through gamification. The VR games include Pro-kin system Immersive VR, VR dodgeball training, VR-assisted CPM, Xbox Kinect, Nintendo Wii, and AR-based platforms, which are used for engagement. VR is recommended as an adjunct to conventional training to enhance effectiveness. Dosage of VR ranges from 20–40 min, 2–3 sessions per week for 6–8 weeks. VR captures accurate kinematic and kinetic movement data, including joint angles, movement patterns, pressure, and velocity. This information supports monitoring and tracking of training and intervention progress to inform further clinical decisions.

According to the findings of an umbrella review based on evidence synthesised from 14 meta-analyses involving over 13,000 patients, VR reduced knee, back, and pain in chronic conditions like fibromyalgia. It also improved functional performance by reducing fatigue, anxiety, and depression, and by improving balance and quality of life. However, mixed results were found in a few conditions, such as neck pain and arthroplasty, with inconsistent findings across various outcomes (Tang et al., 2025).

The effectiveness of VR varies by condition and outcome domain, underscoring the importance of tailoring the application of gaming technologies in MSK rehabilitation. Also, there is still a lack of standardised intervention guidelines for VR and AR in MSK rehabilitation protocols. VR and AR platforms are safe, feasible, home-based, and cost-effective for improving physical activity levels. Overall, gamification through VR platforms increases motivation, ensures strong compliance and adherence to training, improves quality of life by reducing pain intensity, and enhances community participation (Plavoukou et al., 2025).

3.7.6   TelerRehabilitation (TR)

Telerehabilitation (TR) is a safe, effective, and flexible mode of rehabilitation, delivered through information and communication technologies to patients suffering pain, impaired mobility, and function due to chronic MSDs. The spectrum of interventions is broad, covering technologies (telephone, mHealth app, web-based platforms, email, and videoconferencing) and delivery modes (synchronous, asynchronous, and hybrid). It extends clinical services to home/remote settings by conducting assessments, providing education, and providing monitoring and management at a distance. Several systematic reviews have reported the superior effects of TR on musculoskeletal pain and function, as well as on QOL and the psychological domain, compared with usual care or no intervention (Baroni et al., 2023).

In the literature, there is uncertainty regarding the effectiveness of TR compared with conventional treatment for chronic LBP. A very recent SR & MA, synthesising results from 8 RCTs, has found no significant findings on pain and disability in favour of TR. However, they stated that it is a viable alternative for chronic LBP, comparable to conventional treatment. Large trials in large clinical settings, with standardisation of the intervention protocol, are needed to confirm the findings. The common dosage pattern in these studies, as per the FITT protocol, is as follows: moderate-intensity strengthening and stabilisation exercises, lumbar mobility exercises, motor control retraining, and integrated educational modules, applied for 30–60 min, 2–3 sessions/ week for 6–12 weeks (Alahmri et al., 2026). Sensor-integrated tele- or remote rehabilitation with virtual reality interventions and biofeedback has demonstrated multidimensional benefits for chronic LBP, including clinically meaningful pain reduction, improved movement biomechanics, and improved quality of life. These effects are linked to high adherence during sessions when interactivity and biofeedback are embedded within the same protocol. These technologies extend traditional care by offering real-time movement analysis, tailored feedback, and gamified engagement strategies that target all domains of musculoskeletal recovery mechanisms. However, evidence remains heterogeneous, and further long-term studies are needed to confirm sustainability and cost-effectiveness (Garofano et al., 2025). However, the existing evidence is insufficient to conclude the sustainability of the TR intervention program; further studies are recommended to explore the cost-effectiveness and long-term sustainability of TR in MSD (Garofano et al., 2025).

TR is considered an appropriate and outcome-oriented approach, along with conventional rehabilitation of patients with chronic neck pain, to reduce the impact of physical disability. The literature also supports the TR effectiveness in pain reduction, but significant findings on promoting functional mobility and reducing the impairment associated with disability. The TR platform provides synchronous and asynchronous modes for effective delivery, including mobile apps, web-based applications, and videoconferencing. These approaches have a substantial impact on the provision of care to populations living in remote areas.

The findings of the SR and MA reports indicate that the TR approach is beneficial for physical disability associated with knee OA, reducing pain and improving functional mobility. The study evaluated 734 patients recruited across 6 RCTs, in which home-based rehabilitation was superior to conventional interventions. Home-based intervention also reduces travel costs, financial constraints, and the overall economic burden on the family and society. WOMAC was used to measure physical function and pain. TR is considered a suitable approach for pain relief but is not equally effective at improving physical function. Health care was provided through different interventional protocols, patient education modules, exercises, and psychological support for health-related advice and the management of psychological issues. Further studies with long-term evaluation and objective-oriented findings are required to explore the impact of TR (Xiang et al., 2023).

3.8 Future Directions in MSK Technological Rehabilitation

Technological innovation is redefining musculoskeletal rehabilitation, yet its success will ultimately be measured by its ability to improve meaningful functional recovery. The future of technology-based MSK rehabilitation lies in the application and integration of these systems into clinical management. Technologies such as artificial intelligence (AI), machine learning (ML), robotics, wearable sensors, and digital therapeutics have shown promising results in MSK assessment and rehabilitation. However, future research must focus on improving long-term functional outcomes rather than depending upon short-term performance outcome measures. There is also a need to integrate data from patients' clinical findings, biomechanical measures, imaging, and wearable devices to develop a holistic management plan. The future of MSK rehabilitation is not the replacement of therapists, but about strengthening their role with the support of innovative technologies.

References

Abdullah, M., Hulleck, A. A., Katmah, R., Khalaf, K., & El-Rich, M. (2024). Multibody dynamics-based musculoskeletal modeling for gait analysis: A systematic review. *Journal of Neuroengineering and Rehabilitation, 21*(1), 178.

Al Dibs, T., Nassar, N. A. A., Atwa, Z., & Hazboun, S. (2025). AI-powered posture assessment and physiotherapy recommendation using computer vision and deep learning. *Procedia Computer Science, 272*, 522–527.

Alahmri, F., Nuhmani, S., & Muaidi, Q. (2026). Effectiveness of telerehabilitation on chronic low back pain: Systematic review and meta-analysis. *International Journal of Medical Informatics, 206*, 106174. https://doi.org/10.1016/j.ijmedinf.2025.106174

Aleksić, J. (2023). Computer vision solutions for range of motion assessment. *Southeastern European Medical Journal, 7*(1), 55–66. https://doi.org/10.26332/seemedj.v7i1.276

Amorim, P., Paulo, J. R., Silva, P. A., Peixoto, P., Castelo-Branco, M., & Martins, H. (2021). Machine learning applied to low back pain rehabilitation–a systematic review. *International Journal of Digital Health, 1*(1), 10.

Banyai, A. D., & Brişan, C. (2024). *Robotics in physical rehabilitation: Systematic review., 12*(17), 1720.

Baroni, M. P., Jacob, M. F. A., Rios, W. R., Fandim, J. V., Fernandes, L. G., Chaves, P. I., Fioratti, I., & Saragiotto, B. T. (2023). The state of the art in telerehabilitation for musculoskeletal conditions. *Archives of Physiotherapy, 13*(1), 1.

Benatti de Oliveira, G., Vilar Fernandes, L., Amaral, T. F., Vasques, A. C. J., & Pires Corona, L. (2025). Validity and reliability of Gripwise digital dynamometer in the assessment of handgrip strength in older adults. *Frontiers in Aging, 6*, 1560097. https://doi.org/10.3389/fragi.2025.156 0097

Bernhardsson, S., Larsson, A., Bergenheim, A., Ho-Henriksson, C.-M., Ekhammar, A., Lange, E., Larsson, M. E. H., Nordeman, L., Samsson, K. S., & Bornhöft, L. (2023). Digital physiotherapy assessment vs conventional face-to-face physiotherapy assessment of patients with musculoskeletal disorders: A systematic review. *PLoS One, 18*(3), e0283013. https://doi.org/10.1371/journal.pone.0283013

Bouziri, H., Roquelaure, Y., Descatha, A., Dab, W., & Jean, K. (2023). Temporal and spatial distribution of musculoskeletal disorders from 1990 to 2019: A systematic analysis of the global burden of disease. *BMJ. Public Health, 1*(1). https://doi.org/10.1136/bmjph-2023-000353

Buckwalter, J. A. (1996). Effects of early motion on healing of musculoskeletal tissues. *Hand Clinics, 12*(1), 13–24.

Cakmak, A. (2023). Adaptation of the musculoskeletal system to exercise. In D. Kaya Utlu (Ed.), *Functional exercise anatomy and physiology for physiotherapists* (pp. 373–389). Springer International Publishing. https://doi.org/10.1007/978-3-031-27184-7_18

Cao, F., Xu, Z., Li, X.-X., Fu, Z.-Y., Han, R.-Y., Zhang, J.-L., Wang, P., Hou, S., & Pan, H.-F. (2024). Trends and cross-country inequalities in the global burden of osteoarthritis, 1990–2019: A population-based study. *Ageing Research Reviews, 99*, 102382. https://doi.org/10.1016/j.arr. 2024.102382

Castelli, L., Iacovelli, C., Ciccone, S., Geracitano, V., Loreti, C., Fusco, A., Biscotti, L., Padua, L., & Giovannini, S. (2023a). Robotic-assisted rehabilitation of lower limbs for orthopedic patients (roar-o): A randomized controlled trial. *Applied Sciences, 13*(24), 13208.

Castelli, L., Iacovelli, C., Ciccone, S., Geracitano, V., Loreti, C., Fusco, A., Biscotti, L., Padua, L., & Giovannini, S. (2023b). RObotic-assisted rehabilitation of lower limbs for orthopedic patients (ROAR-O): A randomized controlled trial. *Applied Sciences, 13*(24), 13208. https://doi.org/10. 3390/app132413208

Chen, N., Fong, D. Y. T., & Wong, J. Y. H. (2023). Health and economic outcomes associated with musculoskeletal disorders attributable to high body mass index in 192 countries and territories in 2019. *JAMA Network Open, 6*(1), e2250674. https://doi.org/10.1001/jamanetworkopen.2022. 50674

Chen, P., Prosser, M., Phillips, B., & Ellison CEng, P., & Rangan, A. (2025). Accuracy and reliability of remote shoulder motion capturing methods: A systematic review and meta-analysis. *Journal of Shoulder and Elbow Surgery, S1058-2746*(25), 00849–00843. https://doi.org/10.1016/j.jse. 2025.11.007

Chen, S., Chen, M., Wu, X., Lin, S., Tao, C., Cao, H., Shao, Z., & Xiao, G. (2022). Global, regional and national burden of low back pain 1990–2019: A systematic analysis of the global burden of disease study 2019. *Journal of Orthopaedic Translation, Signaling Interaction during OA Development, 32*, 49–58. https://doi.org/10.1016/j.jot.2021.07.005

Cieza, A., Causey, K., Kamenov, K., Hanson, S. W., Chatterji, S., & Vos, T. (2021). Global estimates of the need for rehabilitation based on the global burden of disease study 2019: A systematic analysis for the global burden of disease study 2019. *Lancet, 396*(10267), 2006–2017. https:// doi.org/10.1016/S0140-6736(20)32340-0 de Melo Santana, B., Moura, J. R., de Toledo, A. M., Burke, T. N., Probst, L. F., Pasinato, F., & Carregaro, R. L. (2023). Efficacy of mHealth interventions for improving the pain and disability of individuals with chronic low back pain: Systematic review and meta-analysis. *JMIR mHealth and uHealth, 11*(1), e48204.

El-Tallawy, S. N., Pergolizzi, J. V., Vasiliu-Feltes, I., Ahmed, R. S., LeQuang, J. K., El-Tallawy, H. N., Varrassi, G., & Nagiub, M. S. (2024). Incorporation of "artificial intelligence" for objective pain assessment: A comprehensive review. *Pain and therapy, 13*(3), 293–317. https://doi.org/10.1007/s40122-024-00584-8

Errabity, A., Calmels, P., Han, W.-S., Bonnaire, R., Pannetier, R., Convert, R., & Molimard, J. (2023). The effect of low back pain on spine kinematics: A systematic review and meta-analysis. *Clinical biomechanics, 108*, 106070. https://doi.org/10.1016/j.clinbiomech.2023.106070

Escorpizo, R. (2014a). Defining the principles of musculoskeletal disability and rehabilitation. *Best Practice & Research. Clinical Rheumatology, 28*(3), 367–375. https://doi.org/10.1016/j.berh.2014.09.001

Escorpizo, R. (2014b). Defining the principles of musculoskeletal disability and rehabilitation. *Best Practice & Research. Clinical Rheumatology, 28*(3), 367–375. https://doi.org/10.1016/j.berh.2014.09.001

Fatoye, F., Gebrye, T., Mbada, C. E., & Useh, U. (2023). Clinical and economic burden of low back pain in low- and middle-income countries: A systematic review. *BMJ Open, 13*(4), e064119. https://doi.org/10.1136/bmjopen-2022-064119

Fejer, R., & Hartvigsen, J. (2008). Neck pain and disability due to neck pain: What is the relation? *European Spine Journal: Official Publication of the European Spine Society, the European Spinal Deformity Society, and the European Section of the Cervical Spine Research Society, 17*(1), 80–88. https://doi.org/10.1007/s00586-007-0521-9

Feng, S., Mäntymäki, M., & Pappas, I. O. (2026). Sleep tracking: An integrative review, conceptual framework and future research agendas. *Behaviour & Information Technology*, 1–31.

Garofano, M., Del Sorbo, R., Calabrese, M., Giordano, M., Di Palo, M. P., Bartolomeo, M., Ragusa, C. M., Ungaro, G., Fimiani, G., Di Spirito, F., Amato, M., Ciccarelli, M., Pascarelli, C., Scanniello, G., Bramanti, P., & Bramanti, A. (2025). Remote rehabilitation and virtual reality interventions using motion sensors for chronic low Back pain: A systematic review of biomechanical, pain, quality of life, and adherence outcomes. *Technologies, 13*(5), 186. https://doi.org/10.3390/technologies13050186

Gill, T. K., Mittinty, M. M., March, L. M., Steinmetz, J. D., Culbreth, G. T., Cross, M., Kopec, J. A., Woolf, A. D., Haile, L. M., Hagins, H., Ong, K. L., Kopansky-Giles, D. R., Dreinhoefer, K. E., Betteridge, N., Abbasian, M., Abbasifard, M., Abedi, K., Adesina, M. A., Aithala, J. P., et al. (2023). Global, regional, and national burden of other musculoskeletal disorders, 1990–2020, and projections to 2050: A systematic analysis of the global burden of disease study 2021. *The Lancet Rheumatology, 5*(11), e670–e682. https://doi.org/10.1016/S2665-9913(23)00232-1

Guan, S.-Y., Zheng, J.-X., Feng, X.-Y., Zhang, S.-X., Xu, S.-Z., Wang, P., Cai, H.-Y., & Pan, H.-F. (2025). The impact of population ageing on musculoskeletal disorders in 204 countries and territories, 1990-2021: Global burden and healthcare costs. *Annals of the Rheumatic Diseases, 84*(12), 2128–2138. https://doi.org/10.1016/j.ard.2025.08.002

Guan, S.-Y., Zheng, J.-X., Sam, N. B., Xu, S., Shuai, Z., & Pan, F. (2023). Global burden and risk factors of musculoskeletal disorders among adolescents and young adults in 204 countries and territories, 1990–2019. *Autoimmunity Reviews, 22*(8), 103361. https://doi.org/10.1016/j.autrev.2023.103361

Han, X., Guffanti, D., & Brunete, A. (2025). A comprehensive review of vision-based sensor systems for human gait analysis. *Sensors, 25*(2), 498.

He, X., Zhou, H., Jiang, Y. S., Liu, D.-C., Qi, F., & Wang, Z. (2025a). Comparison of mobile health-based exercise vs traditional exercise for chronic neck pain: A systematic review and meta-analysis. *Journal of Pain Research*, 4639–4649.

He, X., Zhou, H., Jiang, Y. S., Liu, D.-C., Qi, F., & Wang, Z. (2025b). Comparison of mobile health-based exercise vs traditional exercise for chronic neck pain: A systematic review and meta-analysis. *Journal of Pain Research*, 4639–4649.

Hill, B. G., Eble, S., Moschetti, W. E., & Schilling, P. L. (2025). The discordance between pain and imaging in knee osteoarthritis. *The Journal of the American Academy of Orthopaedic Surgeons, 33*(14), e786–e794. https://doi.org/10.5435/JAAOS-D-24-00509

Huo, J., Yu, Y., Lin, W., Hu, A., & Wu, C. (2024). Application of AI in multilevel pain assessment using facial images: Systematic review and meta-analysis. *Journal of Medical Internet Research, 26*, e51250. https://doi.org/10.2196/51250

Hwang, U., Kim, J., Kim, K., & Chung, K. (2024). Machine learning models for predicting return to sports after anterior cruciate ligament reconstruction: Physical performance in early rehabilitation. *DIGITAL HEALTH, 10*, 20552076241299065.

James, S. L., Abate, D., Abate, K. H., Abay, S. M., Abbafati, C., Abbasi, N., Abbastabar, H., Abd-Allah, F., Abdela, J., Abdelalim, A., Abdollahpour, I., Abdulkader, R. S., Abebe, Z., Abera, S. F., Abil, O. Z., Abraha, H. N., Abu-Raddad, L. J., Abu-Rmeileh, N. M. E., Accrombessi, M. M. K., et al. (2018). Global, regional, and national incidence, prevalence, and years lived with disability for 354 diseases and injuries for 195 countries and territories, 1990–2017: A systematic analysis for the global burden of disease study 2017. *The Lancet, 392*(10159), 1789–1858. https://doi.org/10.1016/S0140-6736(18)32279-7

Kalli, I., Niglas, M., Naeini, M. K., Freidin, M., Thomas, L., Menni, C., & Williams, F. (2025). Paraspinal muscle quality in chronic low back pain: A systematic review and meta-analysis of muscle atrophy and fat infiltration. *European Spine Journal: Official Publication of the European Spine Society, the European Spinal Deformity Society, and the European Section of the Cervical Spine Research Society.* https://doi.org/10.1007/s00586-025-09454-z

Kang, L.-J., Lin, P.-Y., Granlund, M., Chen, C.-L., Sung, W.-H., & Chiu, Y.-L. (2023). Development and usability of an app-based instrument of participation in children with disabilities. *Scandinavian Journal of Occupational Therapy, 30*(3), 322–333.

Kapil, D., Wang, J., Olawade, D. B., & Vanderbloemen, L. (2025). AI-assisted physiotherapy for patients with non-specific low back pain: A systematic review and meta-analysis. *Applied Sciences, 15*(3), 1532.

Kazeminasab, S., Nejadghaderi, S. A., Amiri, P., Pourfathi, H., Araj-Khodaei, M., Sullman, M. J. M., Kolahi, A.-A., & Safiri, S. (2022a). Neck pain: Global epidemiology, trends and risk factors. *BMC Musculoskeletal Disorders, 23*(1), 26. https://doi.org/10.1186/s12891-021-04957-4

Kazeminasab, S., Nejadghaderi, S. A., Amiri, P., Pourfathi, H., Araj-Khodaei, M., Sullman, M. J. M., Kolahi, A.-A., & Safiri, S. (2022b). Neck pain: Global epidemiology, trends and risk factors. *BMC Musculoskeletal Disorders, 23*(1), 26. https://doi.org/10.1186/s12891-021-04957-4

Kiesel, K., Matsel, K., Bullock, G., Arnold, T., & Plisky, P. (2024). Risk factors for musculoskeletal health: A review of the literature and clinical application. *International Journal of Sports Physical Therapy, 19*(10), 1255–1262. https://doi.org/10.26603/001c.123485

Koul, M., Thanki, R., Koul, R., Fujarski, S., Radhakrishnan, S., & Jodbhavi, S. (2025). An AI-driven video based goniometer for knee joint range of motion (ROM) assessment: Reliability and validity compared to traditional goniometry. *Computers in Biology and Medicine, 196*(Pt B), 110848. https://doi.org/10.1016/j.compbiomed.2025.110848

Langworthy, M., Dasa, V., & Spitzer, A. I. (2024). Knee osteoarthritis: Disease burden, available treatments, and emerging options. *Therapeutic Advances in Musculoskeletal Disease, 16*, 1759720X241273009. https://doi.org/10.1177/1759720X241273009

Lee, Y. K., Yoon, E.-J., Kim, T. H., Kim, J.-I., & Kim, J.-H. (2025). Musculoskeletal digital therapeutics and digital health rehabilitation: A global paradigm shift in orthopedic care. *Journal of Clinical Medicine, 14*(23), 8467.

Liu, S., Wang, B., Fan, S., Wang, Y., Zhan, Y., & Ye, D. (2022). Global burden of musculoskeletal disorders and attributable factors in 204 countries and territories: A secondary analysis of the global burden of disease 2019 study. *BMJ Open, 12*(6), e062183. https://doi.org/10.1136/bmjopen-2022-062183

Luo, Z., Wang, Y., Zhang, T., & Wang, J. (2025). Effectiveness of AI-assisted rehabilitation for musculoskeletal disorders: A network meta-analysis of pain, range of motion, and functional outcomes. *Frontiers in Bioengineering and Biotechnology, 13*, 1660524.

Mapinduzi, J., Ndacayisaba, G., Verbrugghe, J., Timmermans, A., Kossi, O., & Bonnechère, B. (2025). Effectiveness of mHealth interventions to improve pain intensity and functional disability in individuals with hip or knee osteoarthritis: A systematic review and meta-analysis. *Archives of Physical Medicine and Rehabilitation, 106*(2), 280–291.

Marín-Méndez, H., Marín-Novoa, P., Jiménez-Marín, S., Isidoro-Garijo, I., Ramos-Martínez, M., Bobadilla, M., Mirpuri, E., & Martínez, A. (2021). Using a robot to treat non-specific low back pain: Results from a two-arm, single-blinded, randomized controlled trial. *Frontiers in Neurorobotics, 15*, 715632.

Melman, A., Lord, H. J., Coombs, D., Zadro, J., Maher, C. G., & Machado, G. C. (2023). Global prevalence of hospital admissions for low back pain: A systematic review with meta-analysis. *BMJ Open, 13*(4), e069517. https://doi.org/10.1136/bmjopen-2022-069517

Misailidou, V., Malliou, P., Beneka, A., Karagiannidis, A., & Godolias, G. (2010). Assessment of patients with neck pain: A review of definitions, selection criteria, and measurement tools. *Journal of Chiropractic Medicine, 9*(2), 49–59. https://doi.org/10.1016/j.jcm.2010.03.002

Mokri, C., Bamdad, M., & Abolghasemi, V. (2022). Muscle force estimation from lower limb EMG signals using novel optimised machine learning techniques. *Medical & Biological Engineering & Computing, 60*(3), 683–699. https://doi.org/10.1007/s11517-021-02466-z

Moran, T. E., Ignozzi, A. J., Burnett, Z., Bodkin, S., Hart, J. M., & Werner, B. C. (2022). Deficits in contralateral limb strength can overestimate limb symmetry index after anterior cruciate ligament reconstruction. *Arthroscopy, Sports Medicine, and Rehabilitation, 4*(5), e1713–e1719. https://doi.org/10.1016/j.asmr.2022.06.018

Peniche Silva, C. J., van Griensven, M., & Joris, V. (2026). The use of molecular cues to regenerate musculoskeletal tissues. *Innovative Surgical Sciences, 11*(1), 83–94. https://doi.org/10.1515/iss-2025-0033

Pires, I. M., Denysyuk, H. V., Villasana, M. V., Sá, J., Marques, D. L., Morgado, J. F., Albuquerque, C., & Zdravevski, E. (2022). Development technologies for the monitoring of six-minute walk test: A systematic review. *Sensors, 22*(2), 581.

Plavoukou, T., Staktopoulos, P., Papagiannis, G., Stasinopoulos, D., & Georgoudis, G. (2025). Virtual and augmented reality for chronic musculoskeletal rehabilitation: A systematic review and exploratory meta-analysis. *Bioengineering, 12*(7), 745. https://doi.org/10.3390/bioengineering12070745

Rani, B., Gupta, M., Ganesh, V., Sharma, R., Bhatia, A., & Ghai, B. (2025). Efficacy of mobile health interventions in the conservative management of chronic low back pain in low-and middle-income countries: A systematic review, meta-analysis, and trial sequential analysis. *Pain Reports, 10*(2), e1242.

Santos, V. M., Gomes, B. B., Neto, M. A., & Amaro, A. M. (2024). A systematic review of insole sensor technology: Recent studies and future directions. *Applied Sciences, 14*(14), 6085. https://doi.org/10.3390/app14146085

Scataglini, S., Abts, E., Van Bocxlaer, C., Van den Bussche, M., Meletani, S., & Truijen, S. (2024). Accuracy, validity, and reliability of Markerless camera-based 3D motion capture systems versus marker-based 3D motion capture Systems in Gait Analysis: A systematic review and meta-analysis. *Sensors, 24*(11), 3686. https://doi.org/10.3390/s24113686

Semiz, B., Hancioglu, Ö. K., & Şahin, R. S. (2026). Pain assessment and determination methods with wearable sensors: A scoping review. *Medical & Biological Engineering & Computing, 64*(1), 9–26. https://doi.org/10.1007/s11517-025-03448-1

Shepherd, J., Hansjee, S., Divall, P., Raval, P., & Singh, H. P. (2024). How do digital range of motion measurement devices "measure-up" to traditional goniometry in assessing shoulder range of motion? A systematic review and meta-analysis. *Shoulder & Elbow, 16*(4), 363–381. https://doi.org/10.1177/17585732231195554

Shimizu, H., Shimoura, K., Iijima, H., Suzuki, Y., & Aoyama, T. (2022). Functional manifestations of early knee osteoarthritis: A systematic review and meta-analysis. *Clinical Rheumatology, 41*(9), 2625–2634. https://doi.org/10.1007/s10067-022-06150-x

Sumner, J., Lim, H. W., Chong, L. S., Bundele, A., Mukhopadhyay, A., & Kayambu, G. (2023). Artificial intelligence in physical rehabilitation: A systematic review. *Artificial Intelligence in Medicine, 146*, 102693.

Sun, P., Li, K., Yao, X., Wu, Z., & Yang, Y. (2023). Association between functional disability with postural balance among patients with chronic low back pain. *Frontiers in Neurology, 14.* https://doi.org/10.3389/fneur.2023.1136137

Tang, L., Wang, M.-M., Wang, H.-X., He, X.-Y., & Jiang, Y.-S. (2025). mHealth-based exercise vs. traditional exercise on pain, functional disability, and quality of life in patients with knee osteoarthritis: A systematic review and meta-analysis of randomized controlled trials. *Frontiers in Physiology, 15*, 1511199.

Tang, P., Cao, Y., Vithran, D. T. A. V., Xiao, W., Wen, T., Liu, S., & Li, Y. (2025). The efficacy of virtual reality on the rehabilitation of musculoskeletal diseases: Umbrella review. *Journal of Medical Internet Research, 27*, e64576. https://doi.org/10.2196/64576

Tanik, F., & Ozer Kaya, D. (2024). Relationships between function, pain severity and psychological and cognitive levels in people with chronic neck pain: Cross-sectional study. *Pain Management Nursing, 25*(6), 645–651. https://doi.org/10.1016/j.pmn.2024.06.008

Uivaraseanu, B., Vesa, C. M., Tit, D. M., Abid, A., Maghiar, O., Maghiar, T. A., Hozan, C., Nechifor, A. C., Behl, T., Patrascu, J. M., & Bungau, S. (2022). Therapeutic approaches in the management of knee osteoarthritis (review). *Experimental and Therapeutic Medicine, 23*(5), 328. https://doi.org/10.3892/etm.2022.11257

Urits, I., Burshtein, A., Sharma, M., Testa, L., Gold, P. A., Orhurhu, V., Viswanath, O., Jones, M. R., Sidransky, M. A., Spektor, B., & Kaye, A. D. (2019). Low Back pain, a comprehensive review: Pathophysiology, diagnosis, and treatment. *Current Pain and Headache Reports, 23*(3), 23. https://doi.org/10.1007/s11916-019-0757-1

Wang, Q., Guan, H., Wang, C., Lei, P., Sheng, H., Bi, H., Hu, J., Guo, C., Mao, Y., Yuan, J., Shao, M., Jin, Z., Li, J., & Lan, W. (2025). A wireless, self-powered smart insole for gait monitoring and recognition via nonlinear synergistic pressure sensing. *Science. Advances, 11*(16), eadu1598. https://doi.org/10.1126/sciadv.adu1598

World Health Organization. (2011). *World report on disability* [global report]. *World Health Organization..* https://www.who.int/teams/noncommunicable-diseases/sensory-functions-disability-and-rehabilitation/world-report-on-disability

World Health Organization. (2022). Musculoskeletal health. *World Health Organization..* https://www.who.int/news-room/fact-sheets/detail/musculoskeletal-conditions

Xiang, W., Wang, J.-Y., Ji, B.-J., Li, L.-J., & Xiang, H. (2023). Effectiveness of different Telerehabilitation strategies on pain and physical function in patients with knee osteoarthritis: Systematic review and meta-analysis. *Journal of Medical Internet Research, 25*, e40735. https://doi.org/10.2196/40735

Xie, X., Wang, H., Gao, X., Chen, H., & Zhou, L. (2025). *Efficacy of mHealth in patients with chronic neck pain: A systematic review and meta-analysis.* Pain Management Nursing.

Zhang, B.-F., Liu, L., Xu, S.-L., & Yang, Z. (2025). Global, regional and national burden of knee osteoarthritis 1990–2021: A systematic analysis of the global burden of disease study 2021. *Archives of Gerontology and Geriatrics, 136*, 105867. https://doi.org/10.1016/j.archger.2025.105867

Zhang, M., Zhu, L., Lin, S.-Y., Herr, K., Chi, C.-L., Demir, I., Dunn Lopez, K., & Chi, N.-C. (2023a). Using artificial intelligence to improve pain assessment and pain management: A scoping review. *Journal of the American Medical Informatics Association: JAMIA, 30*(3), 570–587. https://doi.org/10.1093/jamia/ocac231

Zhang, M., Zhu, L., Lin, S.-Y., Herr, K., Chi, C.-L., Demir, I., Dunn Lopez, K., & Chi, N.-C. (2023b). Using artificial intelligence to improve pain assessment and pain management: A scoping review. *Journal of the American Medical Informatics Association, 30*(3), 570–587.

Chapter 4
Technological Interventions for Paediatric Motor Disabilities

Arshad Nawaz Malik and Suman Sheraz

Abstract Developmental and paediatric disabilities affect a significant number of children worldwide. Paediatric disabilities have a serious psychological and functional impact on the quality of life of children. Early intervention is crucial for achieving recovery and for promoting normal developmental processes. This chapter provides an overview of disabilities, including a wide range of conditions such as Cerebral Palsy, Spina Bifida, Muscular Dystrophies, Developmental Delay, Global Developmental Delay, and Juvenile Idiopathic Arthritis. The chapter explains how technology-assisted rehabilitation significantly impacts children's recovery and performance through repetitive, engaging, and outcome-oriented activities grounded in the principles of neural plasticity and motor relearning. It highlights the innovations in technology, including robot-assisted training, virtual & augmented reality gamification platforms, wearable sensors, and Telerehabilitation platforms. The chapter also mentions the few challenges linked with these technologies and highlights the future direction of emerging technology to achieve maximum functional recovery in children.

Keywords Child development · Early intervention · Motor function · Neurodevelopmental disorders · Paediatric rehabilitation

4.1 Overview of Developmental and Paediatric Disabilities

Developmental and paediatric disabilities are a group of conditions that arise during prenatal, natal, early childhood, or adolescent development and interfere with a child's physical, cognitive, emotional, or social development. These conditions may

A. N. Malik · S. Sheraz (✉)
Faculty of Rehabilitation and Allied Health Sciences, Riphah International University, Islamabad, Pakistan
e-mail: suman.sheraz@riphah.edu.pk

A. N. Malik
e-mail: arshad.nawaz@riphah.edu.pk

originate from neurodevelopmental disturbances, congenital structural anomalies, genetic and neuromuscular disorders, or paediatric chronic inflammatory and musculoskeletal diseases. There is a growing global burden of paediatric and developmental disabilities, with an estimated 240 million children and adolescents living with one or more developmental disabilities, accounting for approximately one in every ten children. An estimated 53 million children under the age of 5 years are living with developmental disabilities worldwide and need rehabilitation interventions for their health and wellness. Children with physical disabilities represent a global health challenge during this period of rapid growth and neurodevelopment. The United Nations Sustainable Development Goals (SDGs) have prioritised the health and well-being of populations by 2030, especially in Low-and Middle-Income Countries (LMICs). According to the International Classification of Functioning, Disability, and Health (ICF) framework, developmental disabilities should not be viewed solely as structural and functional impairments, but also as limitations in activity and restrictions in participation influenced by environmental and personal factors.

4.1.1 Definition and Etiological Classification

The Convention on the Rights of Persons with Disabilities (CRPD) define the person with disability as "Those who have long-term physical, mental, intellectual, or sensory impairments which in interaction with various barriers may hinder their full and effective participation in society on an equal basis with others". Developmental disabilities are defined as "Chronic physical, cognitive, speech or language, psychological, or self-care conditions that typically originate during childhood before the age of 22 years" (Olusanya et al., 2023). Based on their aetiology, developmental and paediatric disabilities can be broadly classified into the following categories:

Neurodevelopmental Disorders are a group of conditions resulting from disturbances in brain development during the prenatal, perinatal, or early postnatal period. This leads to impairments in children's motor, cognitive, and sensory functions, affecting their mobility. Physical disabilities in these children often manifest as abnormal muscle tone, poor coordination, delayed motor milestones, and limited mobility, restricting participation in daily activities. Examples include Cerebral Palsy, Global Developmental Delay, Developmental Coordination Disorder (DCD), and early Childhood Traumatic Brain Injury.

Congenital Structural Conditions are anatomical anomalies present at birth that cause musculoskeletal impairments ranging from mild motor difficulties to severe mobility impairments. They can cause limitations in posture, gait, and overall physical activity, often requiring early orthotic or surgical management combined with rehabilitation. Examples include Spina Bifida, Arthrogryposis Multiplex Congenita, Congenital Limb Deficiencies, Clubfoot, and Congenital Hip Dislocation.

Genetic and Neuromuscular Disorders involve inherited or spontaneous genetic mutations that impair muscle strength, coordination, or nerve function. Physical disabilities commonly manifest as progressive weakness, reduced endurance, and impaired mobility, which may severely limit independent function. Conditions included under this are Muscular Dystrophies (Duchenne, Becker), Spinal Muscular Atrophy (SMA), Congenital Myopathies, and Charcot–Marie–Tooth disease.

Chronic Inflammatory and Musculoskeletal Conditions develop during childhood and involve chronic inflammation of joints, connective tissue, or musculoskeletal structures, leading to pain, stiffness, and limited physical function. Physical disabilities may include a restricted range of motion, reduced strength, and impaired endurance, which can affect participation in daily activities and school-based tasks. These conditions include Juvenile Idiopathic Arthritis, Juvenile Dermatomyositis, Paediatric Chronic Pain Syndromes, and Scheuermann's Kyphosis.

Physical disabilities in children manifest as impairments in mobility, strength, balance, coordination, and endurance across all developmental and paediatric disabilities. All these impairments can limit sitting, standing, and mobility, as well as balance and coordination, leading to activity limitations. These mobility limitations also hinder children's social participation, ultimately affecting their development and learning in early childhood.

4.1.2 Functional and Psychosocial Impact of Developmental Disabilities

Children with developmental and paediatric disabilities have significant limitations of function within the domains of physical, cognitive, and social health. The children have limitations in physical function owing to reduced muscle strength, abnormal tone, poor coordination, delayed motor milestones, and impaired mobility. This restricts their participation in self-care, play, and school activities. Cognitive impairments include attention deficit, difficulties with executive functioning, poor problem-solving, and learning. These children also have limited social participation and face challenges in communication and interaction with their peers. These impairments lead to limited interaction with the environment, affecting the acquisition of motor skills, cognitive development, and participation in the community and education.

The consequences of these developmental and paediatric disabilities are not just limited to the functional limitation, but also affect their developmental milestones, age-appropriate skill development, and emotional well-being. These disabilities are often associated with mobility limitations, motor impairments, delayed development, limited activity, and restricted participation in socialisation, play, education, and community activities. Such impairments affect normal growth and productivity in children during their formative years. These disabilities also affect them psychosocially, where they have low self-esteem, more anxiety, experience social isolation, and behavioural challenges. Children with disabilities also become stigmatised and linked with negative beliefs and behaviours that also increase the risk of violence, abuse, exploitation, and neglect. There is an increased socioeconomic burden due to the greater need and utilisation of healthcare services. The children need long-term care requirements in medical, rehabilitation and educational domains of healthcare, accompanied by caregivers and family stress (Olusanya et al., 2022a).

4.1.3 Importance of Early Intervention and Rehabilitation

Rehabilitation is crucial for children with disabilities, as this is a very critical age, and lacking appropriate rehabilitation intervention can result in long-term disability and declining quality of life. These disabilities lead to functional limitations, including delayed motor milestones, loss of motor control, limited mobility, poor postural control, and restricted participation in the community. There is a service gap and a need for rehabilitation in resource-limited countries and settings. Early rehabilitation is vital for rapid recovery through neural plasticity and motor learning & relearning (Luna Lorente et al., 2024). Early, appropriate intervention prevents secondary complications such as contractures or incorrect movement patterns, thereby reducing the consequences and severity of disabilities among children. It also facilitates smooth transitions from home to school, promoting family involvement, ultimately improving functional independence and quality of life. A structured rehabilitation program is important to address the physical, cognitive, emotional, and social needs of children with developmental disabilities (Olusanya et al., 2022b).

4.2 Mechanisms of Recovery in the Developing Brain

The unique properties of the developing brain strongly influence recovery in children with developmental disabilities. A significant feature of the brain is plasticity, which refers to the anatomical, structural, and functional reorganisation of neurons. Neural plasticity helps adapt to changes in its surroundings and retain memory links associated with task learning. Plasticity has different mechanisms for brain development in children. Synaptic plasticity causes synapses to strengthen or weaken. Synaptic plasticity forms the excessive number of synapses during early development, and Long-Term Potentiation (LTP) and Long-Term Depression (LTD) develop motor and cognitive neural circuits. Structural plasticity creates new neuronal connections via axonal sprouting and generates collateral pathways for information flow. Functional plasticity assumes that loss of function results from activation of adjacent cortical regions. Experience-dependent plasticity modulates neuronal activity in response to environmental stimuli, primarily by shaping input and information from the surrounding environment. Motor learning in children requires high repetition, sensory feedback, and an overall enriched environment to enhance plasticity. Active synaptogenesis, dynamic neural migration, myelination, increased levels of neurotrophic factors, and experience-based circuit plasticity are key concepts. In addition to other mechanisms, biomarkers also play a significant role in neural plasticity; these include

Brain-Derived Neurotrophic Factor (BDNF), growth factor signalling, the GABA system, and modulation of the excitatory-inhibitory balance to promote recovery.

After early brain injury, cortical neuronal circuits replace the damaged areas, and the opposite hemisphere supports reorganisation. Axonal sprouting helps create new structural pathways to restore function to lost circuits. However, these structural changes, or neuronal rewiring, depend on the principle of "use it or lose it". Post-injury, in children, primary motor cortex areas have been disturbed, so secondary areas, the premotor cortex and supplementary motor areas, take on the functional role. Bilateral cortical activation has been observed, and compensation is achieved through the recruitment of additional neural circuits. The difference in recovery includes greater structural plasticity and active synaptogenesis in the child's brain compared to the adult brain. However, the risk of malalignments in children is also higher than in adults, where network connections are stable and mature.

There are a few additional limitations of plasticity in children, including the formation of abnormal networks, epileptogenic, maladaptive motor patterns, and congenital developmental delays. Plasticity can be adaptive or maladaptive; however, the timing, duration, and task-specificity of training determine whether it is so. Remapping of motor cortical activity happens in response to task-specific training. However, maladaptive plasticity can also occur, such as mirror movements in CP, in which the contralateral hand involuntarily imitates movements. Activity-dependent refinement during development creates an imbalance, with the unaffected side dominating the cortical representation, leading the patient to use the unaffected side, as in hemiplegic CP.

Early goal-oriented rehabilitation provides guidance and direction for cortical reorganisation in a normal pattern. Rehabilitation through a structured, innovative, and augmented environment promotes motor recovery. Intensive, task-oriented, and a higher number of repetitions strengthen the structural and functional reorganisation of synapses. The interaction between developmental plasticity, such as experience-dependence, and regenerative biological mechanisms drives brain maturation and recovery. Recovery of the developing brain depends on the stage of development, the severity and nature of injury, the regenerative biological response, and environmental input. The recovery from paediatric brain injury is associated with the interactions among growth, reorganisation, and the brain's repair mechanisms. The recovery process in children is not compensation or restoration of function, but rather a developmental process in altered environments.

The recovery of the brain after injury differs between children and adults. In adults, it primarily focuses on compensating for and restoring function within the existing neural network. In contrast, in children, this recovery is an ongoing developmental process within the affected and developing brain. Children have greater neural plasticity, but they are also particularly vulnerable to developmental disturbances, maladaptation, and delayed deficits. Hence, rehabilitation of children should not only focus on function; a holistic approach should guide the developmental process. These mechanisms together form the scientific basis for rehabilitation strategies that emphasise repetition, feedback, progressive challenge, and meaningful engagement (Costanzo et al., 2023; Johnston, 2009).

4.3 Limitations of Conventional Paediatric Rehabilitation

Different evidence-based conventional rehabilitation approaches for children with paediatric and developmental disabilities are currently in practice. Bobath is a neurodevelopmental approach that aims to improve specific developmental stages and postural control through facilitation and sensory feedback. Motor learning focuses on task-specific training, which is beneficial in neurological paediatric dysfunction. The Margaret Rood approach improves motor function and recovery by stimulating superficial and deep sensation. The Kabat-Knott-Voss technique uses diagonal and functional movement patterns to activate motor function and achieve normal movement. Constraint-Induced Movement Therapy (CIMT) focuses on the affected extremity in hemiplegic CP. Additionally, strength training, stretching, aerobic activities, sensory integration, and hydrotherapy are widely used to manage impairments associated with paediatric disabilities. Rehabilitation approaches are effective in improving postural control, balance, gait patterns, upper-limb function, mobility, and overall quality of life. Children with physical disabilities benefit from the implementation of traditional approaches and can perform their functions independently. However, traditional rehabilitation approaches have limitations. The limitations include a short session duration and less intensive training; 1–3 sessions per week are advised. To achieve neural plasticity, highly repetitive, intensive, longer-duration, and home-based training is required.

Traditional approaches are therapist-dependent on effective training; however, limited staff, access constraints in remote areas, and resource shortages directly affect the recovery process. There is a lack of a standardised intervention protocol for all children, as these training courses are based on therapists' expertise, leading to variation in outcomes. Long-term training protocols are highly recommended for children with physical disabilities, but traditional approaches focus on short-term goals and plans. A lack of active engagement and motivation is a significant challenge in conventional training for long-term follow-up planning. Real-time feedback is a key feature of motor recovery that is also not possible in conventional protocols. Accessibility, travel costs, resource constraints, and limited intensity are barriers to the effective implementation of training. There is also a lack of progress monitoring, objective-oriented assessment, a therapist-centred approach, proper dosage of training, and consistency with real-world task performance. Rehabilitation of paediatric patients is considered stressful, with a longer duration and without motivational intervention. The effective training protocol requires a sensorimotor and cognitive learning model with task-specific, repetitive, and engaging components, which are essential for comprehensive rehabilitation. (Iosa et al., 2022; Srushti Sudhir & Sharath, 2023).

4.4 Rationale for Integrating Emerging Technologies

There is a need to overcome the challenges of traditional rehabilitation approaches to make rehabilitation sessions more engaging for children. Technology-assisted rehabilitation can increase therapy intensity through repetition, motivation, and facilitation through home-based therapy sessions. In LMICs, the prevalence of PD is high, and there is limited access to rehabilitation services and assistive technologies. This higher prevalence demands accessible, cost-effective, and technology-assisted therapy for all children with disabilities. Moreover, early rehabilitation is important for children at different stages of development. This is important because their gait pattern and motor abilities remain highly adaptable, and early rehabilitation can minimise the risk of progression to more severe disability.

Neural plasticity, motor learning, and relearning principles, including repetition, task specification, feedback, motivation, and intensity, lay the foundation for rehabilitation in children. Technology-assisted therapy plays a crucial role in providing intensive, task-specific, and repetitive therapy. These technology-based platforms are not a complete replacement for therapist presence; rather, they assist the therapist in the effective delivery of the rehabilitation program. Emerging technologies, including robotics, gamification, virtual and augmented reality, wearable sensors, and telerehabilitation platforms, are now widely used in the rehabilitation of children with physical disabilities. These technologies provide a conducive, enriching environment through gamification and an interactive, engaging nature, increasing adherence, follow-up, and active participation. Emerging technologies provide customised, real-time feedback, objective-oriented assessment, monitoring, and tracking of long-term plans for effective training strategies. Technology-assisted and child-centred approaches are an urgent need to promote activity, functional recovery, and participation outcomes in children with disabilities (Smythe et al., 2024).

4.5 Emerging Technological Interventions in Paediatric Rehabilitation

Recent advances in paediatric rehabilitation have incorporated various technology-supported interventions that increase exercise intensity, motivation, and engagement in children. By promoting repetitive, task-specific practice and adaptive learning, they support neuroplasticity, motor relearning, and functional recovery in children with developmental and physical disabilities.

4.5.1   Robotic and Assistive Devices

Robot-assisted rehabilitation training is a child-centred technological approach that assists and guides movements of the lower and upper limbs. The robot system provides repetitive movement patterns, controlled activities, and partial-to-full assistance, tailored to the child's needs and abilities. This system manages speed, range of motion, specific training control, and integration with other equipment. A robotic system integrated with a treadmill, a visual platform, or a gamification platform provides real-time feedback to enhance motivation and adherence in children during sessions. Robotic system enables repetitive movement patterns and high-intensity training in a single session, which is not possible with therapist-oriented training. They also provide precise, objective-oriented assessment, feedback, and monitoring of training progress. Robotic devices reduce physical exertion, fatigue, and workload of therapists. They also increase adherence, motivation, and active engagement with the performance of difficult and complex tasks. A robotics system can collect objective data, record performance, gait parameters, and progress, and evaluate outcomes (Cardone et al., 2024).

Three modes of robotics can be used to rehabilitate children: the passive mode, in which there is no active movement, and the robotic system provides passive movement, usually in the early stages. The assisted mode, where the child is unable to complete the task, and the robot assists in task performance to achieve completion. The active or resistive modes provide resistance to improve muscle strength for achieving voluntary control of movement. Currently, AI-based robotics can automatically adjust based on the child's effort, fatigue, movement quality, and force during movement activation. Real-time feedback through visual, auditory, and haptic increases engagement and motor learning during task acquisition (Ezra Tsur & Elkana, 2024a). The robotic system is not meant to replace the therapist, but rather to support the training. Therapists play a significant role in decision-making, adjusting support levels, selecting tasks, and translating the functional capabilities of robotic devices. Children's age, cognitive impairments, and motivation level have a significant impact on outcomes.

4.5.2   Virtual and Augmented-Based Gamified Platforms

Virtual reality (VR), augmented reality (AR), and gamified platforms are increasingly used in paediatric rehabilitation to enhance engagement, motivation, and functional recovery. Gamification integrates game-like elements—rewards, challenges, and levels—into both VR and AR exercises, promoting sustained participation and high repetition. VR creates an interactive environment that promotes high task-specific repetition with immediate feedback, without causing fatigue or boredom in children. VR systems use standard displays with motion capture systems, including the Wii, Kinect, PlayStation, and head-mounted displays (HMDs). Tablet-based games are also commonly used in paediatric training. VR systems interact with the upper and lower extremities and are user-friendly, feasible, cost-effective, and suitable for both clinical and home-based rehabilitation. A VR system provides a virtual environmental experience, focused engagement, real-time feedback, and activities under the supervision of therapists. VR-based gamification is highly effective in improving posture, balance, gross motor function, upper-limb function, and overall task performance. It also enhances children's adherence to and engagement in therapy sessions, promoting long-term compliance. Gamification systems, including Xbox Kinect, Wii, and Nintendo, are commonly used to provide a platform for playing and engaging with a virtual environment. These systems provide real-time feedback, tracking records, and progress monitoring to promote functional recovery and motor learning (Iosa et al., 2022).

VR is an effective approach for both clinical and home-based training to improve motor function, postural control, activity level, and participation when applied in addition to traditional training. VR gaming systems are commercially available, easily accessible, and user-friendly, and they promote family involvement and long-term adherence to help children achieve complex skill control. This approach is also appropriate for providing high-quality training that engages children and encourages active participation. VR provides sensory integration through kinesthetic and proprioceptive sources to enhance equal weight-bearing, promoting symmetry, postural control, and weight shifting. VR training provides a challenging environment for anticipatory and reactive balance control, and real-time feedback enhances movement precision and accuracy, thereby improving task execution quality. VR-based training is safe, cost-effective, and supports home-based training through consistent monitoring and progress recording (Komariah et al., 2024; Maggio et al., 2024).

Augmented Reality (AR) integrates digital elements (images, objects, animations) into the real world, allowing children to interact simultaneously with both physical surroundings and virtual stimuli. AR-based games provide real-time visual feedback and increase exercise repetition through interactive gameplay, thereby maintaining environmental awareness, making them particularly suitable for paediatric populations. AR-based games not only improve physical outcomes but also improve emotional and psychological outcomes by reducing pain and anxiety and promoting emotional comfort and treatment alliance in children. A systematic review that synthesised evidence from 14 studies involving 1057 children to evaluate the effects of AR-based interventions on paediatric patient outcomes. AR interventions were generally found to be feasible, acceptable, and engaging for children. They were associated with reduced procedural pain and anxiety, improved cooperation and emotional comfort, enhanced knowledge acquisition, and greater satisfaction with care (Savaş et al., 2025). By combining immersive VR environments and adaptive AR experiences with gamification, therapy sessions become more enjoyable and effective for children with developmental and physical disabilities.

4.5.3 Neurofeedback and Brain–Computer Interface (BCI)

Neurofeedback is a form of non-invasive Brain–Computer Interface (BCI)-based training. It involves monitoring real-time brain activity, usually via EEG, and providing feedback to the child through visual or auditory cues. The child learns to modulate specific brainwave patterns through repeated practice, promoting neuroplastic changes. The neurofeedback focuses on abnormal cortical patterns during movement and helps children learn to control voluntary movement, improving motor function. Apart from motor skills, neurofeedback also has the potential to improve cognitive and functional outcomes, though the evidence is inconsistent. This approach is always beneficial when combined with traditional physical therapy training to regain functional level and control (Abay et al., 2025). Although evidence of this technology is emerging, it remains inconsistent for motor disabilities, mainly due to variations in intervention protocols.

4.5.4 Wearables and Motion Sensors

Wearable sensors (WS) are trackers that collect data on activities and movement to assess the user's personal health and share it with healthcare professionals for appropriate intervention planning. WS devices include wristbands, accelerometer sensors, and smart garments that are used to record movement, step count, heart rate, respiratory rate, sleep, and mobility level to develop a personalised intervention plan. WS are feasible, user-friendly, and popular among children for tracking real-time performance records without specific clinical-oriented testing (McErlane et al., 2021). The recent advances in WS provide an opportunity to convert innovative technologies to promote motor rehabilitation. WS tracks motion and records quantitative and qualitative analyses of movement patterns, postures, ambulation, and physical activity. Therapists can utilise the outcome of WS to plan the specific, tailored exercise guidance for maximum efficacy in children with physical disabilities.

WS plays a crucial role in early screening, detection, assessment, prevention, and control of long-term diseases. WS manages to provide feedback and guidance on the disease's progression and prognosis to the remote population (González Barral & Servais, 2025). This technology is beneficial for promoting a healthy lifestyle, improving mobility, and addressing the needs of remote populations through an effective strategy (Zhang et al., 2024). In children, small, lightweight, soft, durable, brightly coloured, and gamified designs are used to maximise compliance and long-term utilisation. The in-built program of WS is also different for children, considering age-based normative data for heart rate, respiratory rate, and movement development, and analysing developing milestones to record data precisely. WS in children is commonly engaging, parent-oriented, and aligned with developmental stages, enabling accurate analysis and appropriate disease management planning.

4.5.5 Telerehabilitation Platforms

Telerehabilitation (TR) provides rehabilitation services to remote populations through various information and communication technologies. Synchronous, asynchronous, hybrid, and caregiver-facilitated therapy are modes of TR delivery for children with CP. The technologies used for TR, including video conferencing platforms, web-based therapy, and VR-based sessions, are effective tools for delivering rehabilitation programs. TR is appropriate for integrating daily home activities into meaningful tasks; family member involvement enhances compliance, and game-based activities ensure children's active engagement. TR with goal-specific activities promotes motor learning and neural plasticity, facilitating early, and rapid recovery of function. A recent meta-analysis (2025) has shown that home-based telerehabilitation improves motor outcomes in children with neurological motor disorders. The review suggested telerehabilitation as a reasonable alternative to traditional therapy-based sessions, with improvements in motor function, walking, and small gains in range of motion (Calcaterra et al., 2025a). TR is considered a valuable approach because it enhances accessibility for underserved children, maintains intervention continuity during a pandemic, and is cost-effective and time-saving. TR is a caregiver or family-centred approach that provides an opportunity for self-training with motivation for long-term commitment. However, there is still a lack of standardised protocols and dosage guidelines, limited application in severe impairment, and no objective outcome measures to record progress and monitor training (Calcaterra et al., 2025b; Ogourtsova et al., 2023).

4.6 Disease-Specific Applications of Technological Interventions

The following section highlights selected disease-specific applications of emerging technologies in paediatric rehabilitation.

4.6.1 Cerebral Palsy (CP)

CP is a group of non-progressive chronic disorders of movement and coordination caused by damage in the developing brain. CP is comprised of cognitive impairments, motor dysfunction, communicative and behavioural disturbances. The prevalence of CP is estimated to be 2/1000 live births. Cerebral palsy is the leading paediatric physical disability, presented by posture and movement disorders that reduce functional mobility, activity limitation, and participation restriction. The most disabling impairment in CP is gait dysfunction, which results in reduced walking speed, a short stride length, and limited mobility. Gait training is crucial for families and therapists to help children achieve functional mobility and independence. Gait training is a repetitive, task-specific activity that requires the child's active participation and physical exertion.

Robot-assisted therapy improves gross motor function measure scores, walking capabilities, lower extremity strength, postural control, and mobility in children with moderate impairments. Robot-assisted gait training is an appropriate approach for severe gait impairments, improving mobility and walking performance (Llamas-Ramos et al., 2022). Robot-assisted therapy yields better functional and motor performance outcomes in children with mild-to-moderate impairments than conventional training. According to the literature, robotic-assisted training ranged from 30 to 60 minutes, with 2 to 5 sessions per week for 4 to 12 weeks, for a total of 12 to 40 sessions. However, there is variation in dosage and long-term effectiveness of robot-assisted therapy, and there remains a need for an evidence-based, standardised protocol (Ezra Tsur & Elkana, 2024b; Yang et al., 2024).

A *virtual reality* system creates an environment that enhances visual-motor proprioceptive coordination, and this sensory-visual-motor coupling increases sensor-motor integration, which is commonly impaired in children with CP. VR is effective in improving upper-extremity reaching, grasping, coordination, and posture control, as well as lower-extremity gait parameters, including step length, stride, speed, and symmetry, in children with paediatric and developmental disabilities. It also improves the gross motor function of children with CP and promotes neural plasticity and motor learning through intensive, repetitive, and goal-oriented training. There are variations in protocol dosage, including 20–60-minute sessions, 2–5 sessions per week, and a total duration of 4–12 weeks (12–30 sessions), according to the task-level requirements. Progression and gradually increasing task difficulty are key benefits of VR-based therapy, which allows for hundreds of repetitions to achieve the desired skill. The limitations of VR-based training include limited task transferability to real life, and severe cognitive and visual impairments prevent children from performing tasks (Fang et al., 2025; Kilcioglu et al., 2023). A fully immersive VR system provides a complete virtual environment; these are not frequently used with children due to safety and long-term tolerance concerns. In the paediatric population, there is also a high risk of cybersickness associated with the usage of VR. However, there is potential for active participation, adherence, and engagement in task performance (Wang et al., 2023).

Recently, in CP, the management focus has shifted from impairment-oriented interventions to motor-learning and neural-plasticity-based interventions, using technology and personalised rehabilitation programs. EEG-based sensorimotor neurofeedback demonstrated potential benefits for motor control, upper-limb coordination, movement planning, and gross motor function in children with mild-to-moderate CP. The effects of neurofeedback on spasticity are inconsistent, as there is no direct pathophysiological mechanism to reduce tone; it modulates the tone as a secondary effect. Neurofeedback directly focuses on brain-activity relationships, supports cortical reorganisation, provides interactive, real-time feedback, and promotes recovery through engagement and motivation. The dosage varies from

20–40 minutes, 2–3 sessions per week, 10–30 total sessions in 4–12 weeks of the complete program. The dosage is adjusted based on the level of cognitive impairment, attention span, severity of impairment, and training type (imagery or movement). Neurofeedback is more effective when integrated with the physical practice of tasks rather than applied in isolation.

The *Telerehabilitation* program focuses on activities of daily living, upper-extremity function, and task-specific activities for children with GMFCS levels I-III. TR improves quality of life, enhances motivation, reduces the physical burden of caregivers, and increases adherence to rehabilitation programs. Overall, TR is a better approach, with equal potential and, in some cases, superior to face-to-face therapy. TR is more appropriate for maintaining recovery and sustaining long-term functional retention, especially in children from remote areas. The TR protocol includes a 20–60-minute session duration, 2–5 sessions per week, 4–12 weeks, and a 6-month extension and follow-up program. The protocol's variation enables the therapist to adapt as needed and make other relevant adjustments to resource and accessibility aspects.

4.6.2 Spina Bifida

Spina bifida is a congenital neural tube defect caused by incomplete closure of the spinal cord and vertebral arches during early embryonic development in the first three weeks of pregnancy. During development, the nerves develop a hernia sac outside the spinal cord rather than being embedded within it. This disruption impairs communication from the brain to the extremities and other visceral organs below the level of the lesion. The impairments associated with Spina bifida depend on the severity of the lesion and level of lesion, ranging from mild disruption of sensation to severe paralysis. Common symptoms include paralysis of the lower extremities, sensory deficits, abnormal gait patterns, difficulty ambulating, and bowel and bladder dysfunction. Spina bifida also affects brain development, especially in the midbrain, cerebellum, and corpus callosum. 80% of individuals develop type II Arnold-Chiari malformation, which leads to increased spinal fluid resulting in hydrocephalus. Children with Spina bifida need a comprehensive and long-term rehabilitation care depending upon the level of functional and mobility restrictions. Emerging technologies support the achievement of rehabilitation goals and facilitate functional mobility and independence in children with spina bifida (Lindquist et al., 2022).

Robot-assisted training with a knee exoskeleton provides support and assistance in achieving a normal gait pattern. Loss of knee joint control, decreased gait speed, lower-limb weakness, and reduced walking ability lead to limited activity and social participation. A robotic knee exoskeleton was developed to provide powered support during stance stability, maintaining knee extension during normal walking. Device control has been managed through two strategies: assistive control and training-oriented control. In assistive control, passive knee extension is used, while in training-oriented control, support is based on patient input to promote motor learning. The knee exoskeleton improves walking speed, gait stability, lower-limb mechanics, and overall mobility. The key feature is to shift from passive support to active participation, reflecting the alignment with basic principles of neural plasticity. The short sessions are recommended in a controlled environment for repetitive gait training. The intensity of training is based on fatigue, safety, and ability to manage the task. The protocol of robotic-assisted training includes 30–60-minute sessions, 2–5 sessions per week, and 4–12 weeks of total training, with gradual progression in task difficulty and in accordance with children's tolerance.

VR-based training improves motor symptoms in spina bifida through real-time feedback, activity progression, and an engaging environment. VR provides a platform for enhancing gait speed, walking ability, ambulation capacity, and functional mobility. Children enjoy the training session and participate actively in task-specific activities to improve overall functional capacity (Devine et al., 2025).

Telerehabilitation is a better and more feasible alternative to traditional face-to-face rehabilitation, particularly beneficial for those living in remote areas. TR improves balance, walking ability, and functional status in spina bifida through structured, supervised programs. TR allows us to monitor progress regularly, track training progress, reduce the travel cost burden, and promote family engagement and adherence to training (Μακαρώνη, 2024). WS using inertial and motion-sensor technologies improves gait, balance, and functional ambulation through precise assessment and a specific, customised training protocol. WS provides objective-oriented, precise measurement data over time that guides the therapist in planning and monitoring progress (Bendt et al., 2022).

4.6.3 *Muscular Dystrophies (MD)*

MD are a group of progressive, genetic neuromuscular disorders characterised by degeneration of skeletal muscle fibres and replacement with fat and connective tissue in children. The two common types of MD are Duchenne Muscular Dystrophy (DMD) and Becker Muscular Dystrophy (BMD). The impairments of MD include progressive weakness of proximal muscles, limited functional ambulation, reduced range of motion, early fatigue, and decreased engagement and participation in social activities. MD is a progressive disorder in nature, hence the focus of rehabilitation is not on recovery, but on preserving function, activities, and quality of life. Motor impairments include progressive weakness of proximal muscles, including the hip and shoulder girdles, which initially appear, and children are unable to walk, run, jump, or even stand up from the floor. In later stages, scoliosis develops due to poor trunk control and other postural deformities. MD is also linked with progressive weakness of respiratory muscles, decreased vital capacity, recurrent risks of chest infection, and disturbance of sleep, leading to declining quality of life. These secondary impairments reduce active participation in society, school, and play and increase the burden on family and caregivers. Technology-supported rehabilitation of MD provides a safe, interactive, functional, goal-oriented, and controlled dosage of training to improve the quality of life of children.

VR-based gamification creates a platform for task-oriented functional activities, enhances active engagement, improves postural control during performance, and reduces physical exertion of children. Reduced muscle load enhances the capacity for active movement without early fatigue. Gamification provides an interactive environment, improves adherence to home-based therapy, increases engagement and reduces monotony during sessions, and is therapist-oriented. These trainings improve breathing control and maintain respiratory muscle function, thereby improving cardiopulmonary functional performance. Technology-assisted rehabilitation is cost-effective for long-term implementation (Calderone et al., 2026).

Wearable sensors are also an appropriate technology in MD for tracking, monitoring, supporting personalised training, reducing excessive energy consumption, and enabling long-term follow-up in home settings.

Tele-Rehab platforms enable accessible training delivery in remote areas, support home-based exercise plans, and involve caregiver participation. TR improves adherence to the standard low-intensity protocol for long-term implementation, thereby maintaining functional status in children with MD. TR and other associated digital platforms are widely used to ensure long-term intervention, monitoring, progression, and feedback. The TR offers multiple options for quick, accessible delivery, including synchronous delivery via live video calls, asynchronous delivery via recorded videos, virtual reality-based training, remote respiratory training programs, and digital monitoring of progression. Technologically assisted rehabilitation maintains muscle strength, upper extremity function, ADLs, and functional motor scores, reducing the rate of functional decline and improving ambulation in the early stages. TR is an effective intervention for long-term adherence, increased motivation, and more active caregiver engagement in delivering training to children. TR also provides customised training plans, reduces the need for frequent hospital visits, and supports continuity of long-term care in home settings (Vinolo-Gil et al., 2025).

The dosage protocol should be child-centred and guided by fatigue and the severity of disorders. The typical motor relearning and neuroplasticity protocol is not fully applicable due to fatigue and degenerative states of the musculature. The session should be 15–30 minutes, 2–4 sessions per week, low to moderate intensity, not high-intensity, and long-term application. It is very important to consider the following precautions while applying the technological-assisted rehabilitation plan. The precautions include regular fatigue monitoring, and muscle soreness, pain, and high resistance should be avoided; prioritise a consistent approach, not high intensity, to avoid complications. Emerging technologies provide real-time feedback and adjustments that help maintain training accuracy, unlike therapist-oriented training programs (Kiper et al., 2024).

4.6.4 Developmental Delay

Developmental Delays (DD) is a condition in which a child within the typical age bracket has a delay or significant lag in the development of milestones. The child is unable to achieve age-specific developmental milestones within the expected time-frame compared to peers. DD can be specific to one domain or global, affecting multiple domains of normal development. The delays can occur across different domains, including gross motor, fine motor, language and communication, cognitive, socioemotional, and adaptive self-care. DD can be temporary, constant, specific, global and can be associated with different other conditions, including cerebral palsy, genetic syndrome, prematurity, or environmental deprivation. Motor impairments in DD include delayed neck/head control, in sitting, standing, and walking; postural control; appropriate coordination; upper-extremity function; independence; and overall motor planning. The cognitive impairments include delays in attention, memory formation, problem-solving skills, and higher-order mental function, leading to overall cognitive decline. Development of proper Language skills is crucial for social interaction, play activities, and appropriate communication with peers and family. Such impairments due to developmental delay reduce participation in the community, decline quality of life and limit learning opportunities, so early, accessible and evidence-based holistic interventions are essential for children.

Technology-assisted rehabilitation provides multisensory feedback, and adopting task difficulty levels and increased repetition are important for effective recovery. Emerging technologies are child-centred, play-based platforms that are enjoyable, low-fatiguing, and involve the caregiver, with objective performance monitoring to promote motor learning through neural plasticity. The technologies include virtual reality, game-based apps, mobile-based apps, and a tele-rehab platform for effective training delivery. This shift from therapist-oriented to active engagement of the child is key to focusing on participation and critical development. Technology-assisted rehabilitation sessions last 15–45 minutes, occur 3–5 times per week, and span 6–12 weeks. The main emphasis of the intervention should be on flexible, age-specific, short, interactive, play-oriented sessions, and on avoiding prolonged, monotonous sessions to prevent distraction and fatigue (Del Lucchese et al., 2024; Fu & Ji, 2023).

Motor skills deficiencies include locomotor control (balance and walking), while in object control, impairments include catching and throwing. Physical activity has been reduced in children with DD, which ultimately leads to poor functional status and activity level as compared to peers. VR head-mounted devices provide an immersive, engaging virtual environment, such as a pedalled stationary cycle, that links to physical activity. Physical activity through virtual engagement and real-time feedback improves the functional level, participation, and overall mobility of children with DD. VR enhances children's engagement, interest, motivation, joy, and willingness to participate in gamification activities that benefit motor recovery and learning. VR platforms are safe, controlled, and pose no risk to children during activities. The dosage of VR varies from 30–40 minutes per session, 2–3 days per week and overall,

960 minutes (16 hours) of practice to achieve movement performance (Lee & Jin, 2023).

4.6.5 Global Developmental Delay

Global Developmental Delay (GDD) is a type of DD and a neurodevelopmental condition of children under five years of age, and the prevalence is almost 1–3%. Multiple domains of development characterise it. GDD has a diverse impact on children and different functional impairments that require intensive, early, personalised, and multimodal training strategies to achieve optimal recovery. *VR semi-immersive approach* with the Nirvana system provides engaging games on walls or floors via infrared-based projectors. This system does not require a head-mounted display; rather, children can use their body movements to interact with and react to these virtual games. This system provides real-time visual, auditory, and sensory feedback, reducing cognitive load and enabling greater attention and focus during training. Active participation, high engagement, repetition, and live feedback, along with personalised training, create an enriching environment for neural plasticity and motor learning in children with GDD. The dosage includes 48 total sessions; each session is 40 minutes long, and sessions should be implemented biweekly. *Telerehabilitation* via video conferencing and web-based platforms delivers appropriate training in remote areas. TR ensures highly intensive, repetitive training, goal-specific, feedback-oriented, real-time monitoring, family-centred, and important for neural plasticity and motor learning. TR provides access in remote areas, reduces clinical staff workload, reduces travel costs, minimises waiting lists, and ensures timely care for all children. TR sessions should be 2–5 times per week, 30–60 minutes each session, and 6–12 weeks in accordance with the severity status of children with DD (Ogourtsova et al., 2023).

Mobile applications are used to record developmental milestone achievements in real-time. This provides an observatory and monitoring approach to normal developmental milestones in early childhood. Delay milestone achievement is an indicator of DD and disability in children. Mobile apps accurately record normal milestones, share data in real-time, and monitor early childhood data. These apps detect early delays, facilitate timely referral, and support individual screening, planning, and monitoring of development. Mobile apps are feasible, accessible, reduce parent recall bias, enable longitudinal recording, promote parent engagement, and are cost-effective to implement.

4.6.6 Juvenile Idiopathic Arthritis

Juvenile Idiopathic Arthritis (JIA) is a common rheumatological chronic inflammatory joint disorder that occurs in children. JIA is a heterogeneous group of arthritides of unknown aetiology that occur before the age of 16 years and persist for more than six weeks. The symptoms of JIA include pain, joint inflammation, stiffness, and limited range of motion. The impairments lead to restricted mobility, reduced participation, poor fine-motor skills, and long-term joint deformities. The rehabilitation of JIA primarily focuses on pain management, joint mobility, functional level, and children's mobility and participation. Multiple rehabilitation strategies, including exercise therapy, modalities, orthoses, thermotherapy, cryotherapy, and laser therapy, have been used to improve mobility and activity levels, prevent deformities, and promote quality of life (Chickermane et al., 2023).

Telerehabilitation is considered an interactive platform for patients and families, in addition to routine physical therapy sessions, to enhance compliance with long-term training and follow-up. TR through video conferencing for 25–30-minute sessions to patients from remote areas to improve the functional mobility and quality of life in children with JIA. TR is a feasible, cost-effective, interactive, home-based platform and is recommended for JIA to enhance training adherence (Stavrakidou et al., 2023).

Digital interventions are gradually being incorporated into the management of JIA to reduce pain and improve physical activity. The digital tools include mobile apps, websites, wearable sensors, pain management tools, and other platforms that play a significant role in non-pharmacological management. mHealth apps are used to track pain symptoms, monitor, provide exercise reminders, educate about the disease, and provide self-care tools to support adherence to home-based rehabilitation. Mobile apps encourage self-monitoring of fatigue and pain, enhance patients' autonomy, and improve overall self-intervention, but are largely dependent on patients' self-motivation. Web-based platforms are also used for online awareness, peer support counselling, psychological consultation, and pain education guidelines. WS in JIA provide guidance on monitoring activity level and task performance intensity, real-time feedback, and a customised exercise plan for effective recording (Ren et al., 2025).

4.7   Challenges and Future Directions

The integration of emerging technologies in paediatric rehabilitation is improving outcomes for children with paediatric and developmental disabilities, but it is also facing several challenges. High costs of robotic systems, VR platforms, and wearable devices limit accessibility, particularly in low-and middle-income settings. Issues such as data privacy and security, screen time exposure, and over-reliance on technology also need to be addressed. Most importantly, technology should facilitate, not replace, the therapist's expertise and the child's and family's interaction. According to current evidence, variability in study designs, small sample sizes, and short follow-up periods limit the generalisability of the evidence. Standardised protocols and condition-specific outcome measures are still evolving.

The future of technology-driven paediatric rehabilitation lies in developing more personalised, accessible, and child-centred technologies. Adaptive systems powered by artificial intelligence, integration of wearable sensors for real-time monitoring, and hybrid models should be developed in-clinical and home-based telerehabilitation. Greater emphasis should be placed on participation-based outcomes, long-term functional independence, and family involvement, rather than on impairment-level improvements. Collaborative research involving biomedical engineers, clinicians, educators, and families is essential to developing accessible, evidence-based, and child-centred technological solutions that support and improve children's developmental trajectories.

References

Abay, S. N., Alaerts, K., Dan, B., Monbaliu, E., & Bekteshi, S. (2025). Effectiveness of neurofeedback interventions in cerebral palsy: A systematic review. *medRxiv*, 2025–08.

Bendt, M., Forslund, E. B., Hagman, G., Hultling, C., Seiger, Å., & Franzén, E. (2022). Gait and dynamic balance in adults with spina bifida. *Gait & Posture, 96*, 343–350.

Calcaterra, V., Marin, L., Guardamagna, L., Gatti, A., Rossi, V., Patanè, P., Vandoni, M., & Zuccotti, G. (2025a). Home-based telerehabilitation for pediatric neurological motor disorders: Current trends and future perspectives. A systematic review and meta-analysis. *Digital. Health, 11*, 20552076251357504. https://doi.org/10.1177/20552076251357504

Calcaterra, V., Marin, L., Guardamagna, L., Gatti, A., Rossi, V., Patanè, P., Vandoni, M., & Zuccotti, G. (2025b). Home-based telerehabilitation for pediatric neurological motor disorders: Current trends and future perspectives. A systematic review and meta-analysis. *Digital. Health, 11*, 20552076251357504. https://doi.org/10.1177/20552076251357504

Calderone, A., Latella, D., De Luca, R., Gangemi, A., Impellizzeri, F., De Pasquale, P., Corallo, F., Manuli, A., Quartarone, A., Portaro, S., & Calabrò, R. S. (2026). Virtual horizons: Enhancing rehabilitation of neuromuscular diseases through virtual reality and gamification. *Journal of Neuromuscular Diseases, 13*(1), 94–108. https://doi.org/10.1177/22143602241311194

Cardone, D., Perpetuini, D., Di Nicola, M., Merla, A., Morone, G., Ciancarelli, I., Moretti, A., Gimigliano, F., Cichelli, A., De Flaviis, F., Martino Cinnera, A., & Paolucci, T. (2024). Robot-assisted upper limb therapy for personalized rehabilitation in children with cerebral palsy: A systematic review. *Frontiers in Neurology, 15*, 1499249. https://doi.org/10.3389/fneur.2024.1499249

Chickermane, P. R., Panjikaran, N. D., & Balan, S. (2023). Role of rehabilitation in comprehensive management of Juvenile Idiopathic Arthritis: When and how? *Indian Journal of Rheumatology, 18*(1S). https://doi.org/10.4103/injr.injr_55_22

Costanzo, F., Fucà, E., Menghini, D., & Vicari, S. (2023). Brain recovery in childhood: The interaction between developmental plasticity and regenerative mechanisms. In L. Petrosini (Ed.), *Neurobiological and psychological aspects of brain recovery* (pp. 289–317). Springer International Publishing. https://doi.org/10.1007/978-3-031-24930-3_13

Del Lucchese, B., Parravicini, S., Filogna, S., Mangani, G., Beani, E., Di Lieto, M. C., Bardoni, A., Bertamino, M., Papini, M., & Tacchino, C. (2024). The wide world of technological telerehabilitation for pediatric neurologic and neurodevelopmental disorders–a systematic review. *Frontiers in Public Health, 12*, 1295273.

Devine, T. M., Asante-Otoo, A., Alter, K., Damiano, D. L., & Bulea, T. C. (2025). Robotic knee exoskeletons as assistive and gait training tools in spina bifida: A pilot study showing clinical feasibility of two control strategies. *IEEE Transactions on Neural Systems and Rehabilitation Engineering*.

Ezra Tsur, E., & Elkana, O. (2024a). Intelligent robotics in pediatric cooperative neurorehabilitation: A review. *Robotics, 13*(3), 49. https://doi.org/10.3390/robotics13030049

Ezra Tsur, E., & Elkana, O. (2024b). Intelligent robotics in pediatric cooperative neurorehabilitation: A review. *Robotics, 13*(3), 49.

Fang, E., Guan, H., Du, B., Ma, X., & Ma, L. (2025). Effectiveness of virtual reality for functional disorders in cerebral palsy: An overview of systematic reviews and meta-analyses. *Frontiers in Neurology, 16*, 1582110.

Fu, W., & Ji, C. (2023). Application and effect of virtual reality technology in motor skill intervention for individuals with developmental disabilities: A systematic review. *International Journal of Environmental Research and Public Health, 20*(5), 4619.

González Barral, C., & Servais, L. (2025). Wearable sensors in paediatric neurology. *Developmental Medicine and Child Neurology, 67*(7), 834–853. https://doi.org/10.1111/dmcn.16239

Iosa, M., Verrelli, C. M., Gentile, A. E., Ruggieri, M., & Polizzi, A. (2022). Gaming technology for pediatric neurorehabilitation: A systematic review. *Frontiers in Pediatrics, 10*, 775356.

Johnston, M. V. (2009). Plasticity in the developing brain: Implications for rehabilitation. *Developmental Disabilities Research Reviews, 15*(2), 94–101. https://doi.org/10.1002/ddrr.64

Kilcioglu, S., Schiltz, B., Araneda, R., & Bleyenheuft, Y. (2023). Short-to long-term effects of virtual reality on motor skill learning in children with cerebral palsy: Systematic review and meta-analysis. *JMIR Serious Games, 11*(1), e42067.

Kiper, P., Federico, S., Szczepańska-Gieracha, J., Szary, P., Wrzeciono, A., Mazurek, J., Luque-Moreno, C., Kiper, A., Spagna, M., & Barresi, R. (2024). A systematic review on the application of virtual reality for muscular dystrophy rehabilitation: Motor learning benefits. *Life, 14*(7), 790.

Komariah, M., Amirah, S., Abdurrahman, M. F., Handimulya, M. F. S., Platini, H., Maulana, S., Nugrahani, A. D., Mulyana, A. M., Qadous, S. G., Mediani, H. S., & Mago, A. (2024). Effectivity of virtual reality to improve balance, motor function, activities of daily living, and upper limb function in children with cerebral palsy: A systematic review and meta-analysis. *Therapeutics and Clinical Risk Management, 20*, 95–109. https://doi.org/10.2147/TCRM.S432249

Lee, H. K., & Jin, J. (2023). The effect of a virtual reality exergame on motor skills and physical activity levels of children with a developmental disability. *Research in Developmental Disabilities, 132*, 104386.

Lindquist, B., Jacobsson, H., Strinnholm, M., & Peny-Dahlstrand, M. (2022). A scoping review of cognition in spina bifida and its consequences for activity and participation throughout life. *Acta Paediatrica, 111*(9), 1682–1694.

Llamas-Ramos, R., Sánchez-González, J. L., & Llamas-Ramos, I. (2022). Robotic Systems for the Physiotherapy Treatment of children with cerebral palsy: A systematic review. *International Journal of Environmental Research and Public Health, 19*(9), 5116. https://doi.org/10.3390/ijerph19095116

Luna Lorente, B., Polo Martínez, L., Ruiz Penichet, V. M., & Lozano Pérez, M. D. (2024). *Assistive technologies for children with physical disabilities: A systematic literature review.* 1–10.

Maggio, M. G., Valeri, M. C., De Luca, R., Di Iulio, F., Ciancarelli, I., De Francesco, M., Calabrò, R. S., & Morone, G. (2024). The role of immersive virtual reality interventions in pediatric cerebral palsy: A systematic review across motor and cognitive domains. *Brain Sciences, 14*(5), 490. https://doi.org/10.3390/brainsci14050490

McErlane, F., Davies, E. H., Ollivier, C., Mayhew, A., Anyanwu, O., Harbottle, V., & Donald, A. (2021). Wearable Technologies for Children with chronic illnesses: An exploratory approach. *Therapeutic Innovation & Regulatory Science, 55*(4), 799–806. https://doi.org/10.1007/s43441-021-00278-9

Μακαρώνη, Π.-Α. Γ. (2024). Face to face vs telerehabilitation in the child population: Systematic review.

Ogourtsova, T., Boychuck, Z., O'Donnell, M., Ahmed, S., Osman, G., & Majnemer, A. (2023). Telerehabilitation for children and youth with developmental disabilities and their families: A systematic review. *Physical & Occupational Therapy in Pediatrics, 43*(2), 129–175.

Olusanya, B. O., Halpern, R., Cheung, V. G., Nair, M., Boo, N. Y., Hadders-Algra, M., & Global Research on Developmental Disabilities Collaborators. (2022a). Disability in children: A global problem needing a well-coordinated global action. *BMJ Paediatrics Open, 6*(1), e001397.

Olusanya, B. O., Kancherla, V., Shaheen, A., Ogbo, F. A., & Davis, A. C. (2022b). Global and regional prevalence of disabilities among children and adolescents: Analysis of findings from global health databases. *Frontiers in Public Health, 10*, 977453.

Olusanya, B. O., Smythe, T., Ogbo, F. A., Nair, M., Scher, M., & Davis, A. C. (2023). Global prevalence of developmental disabilities in children and adolescents: A systematic umbrella review. *Frontiers in Public Health, 11*, 1122009.

Ren, Z., Chen, Y., Li, Y., Fan, P., Liu, Z., & Shen, B. (2025). Digital interventions for patients with juvenile idiopathic arthritis: Systematic review and meta-analysis. *JMIR Pediatrics and Parenting, 8*, e65826. https://doi.org/10.2196/65826

Savaş, E. H., Coşkun, A. B., Elmaoğlu, E., Semerci, R., & Şahiner, N. C. (2025). Investigating the effects of augmented reality-based interventions on pediatric patient outcomes in the clinical setting: A systematic review. *Journal of Pediatric Nursing, 85*, 39–47. https://doi.org/10.1016/ j.pedn.2025.07.014

Smythe, T., Scherer, N., Nanyunja, C., Tann, C. J., & Olusanya, B. O. (2024). Strategies for addressing the needs of children with or at risk of developmental disabilities in early childhood by 2030: A systematic umbrella review. *BMC Medicine, 22*(1), 51. https://doi.org/10. 1186/s12916-024-03265-7

Srushti Sudhir, C., & Sharath, H. V. (2023). A brief overview of recent pediatric physical therapy practices and their importance. *Cureus, 15*(10), e47863. https://doi.org/10.7759/cureus.47863

Stavrakidou, M., Trachana, M., Koutsonikoli, A., Spanidou, K., & Hristara-Papadopoulou, A. (2023). The impact of a physiotherapy tele-rehabilitation program on the quality of Care for Children with juvenile idiopathic arthritis. *Mediterranean Journal of Rheumatology, 34*(4), 443–453. https://doi.org/10.31138/mjr.310823.tio

Vinolo-Gil, M. J., Muñoz-Pérez, L., Dominguez-Vera, P. A., García-Campanario, I., Estebánez-Pérez, M. J., & Martín-Valero, R. (2025). Effectiveness of telerehabilitation on motor and respiratory function in duchenne muscular dystrophy: A systematic review and meta-analysis. *Disability and Rehabilitation*, 1–20.

Wang, N., Liu, N., Liu, S., & Gao, Y. (2023). Effects of nonimmersive virtual reality intervention on children with spastic cerebral palsy: A meta-analysis and systematic review. *American Journal of Physical Medicine & Rehabilitation, 102*(12), 1130–1138.

Yang, F.-A., Shih, Y.-C., Lin, L.-F., Peng, C.-W., Lai, C.-H., Liou, T.-H., Escorpizo, R., & Chen, H.-C. (2024). Effect of robotic training on the mobility of children with cerebral palsy: A systematic review and meta-analysis of randomized controlled trials. *Rehabilitation Practice and Science, 2024*(1). https://doi.org/10.6315/3005-3846.2241

Zhang, W., Xiong, K., Zhu, C., Evans, R., Zhou, L., & Podrini, C. (2024). Promoting child and adolescent health through wearable technology: A systematic review. *Digital Health, 10*, 20552076241260507. https://doi.org/10.1177/20552076241260507

Chapter 5
Geriatric Physical Disabilities and Technological Advancements

Suman Sherazⓘ

Abstract There is a rapid increase in the ageing population globally, leading to a rising burden of age-related physical disabilities and frailty. Physiological changes in the body, along with comorbidities, impair older adults' ability to perform daily activities, leading to functional limitations, increased dependence on caregivers, and worsened quality of life. Geriatric rehabilitation is a multifaceted approach that addresses diverse issues and improves the functional status and mobility of older adults. This chapter highlights the integration of innovative technologies in geriatric rehabilitation through robot-assisted interventions, virtual reality, exergaming, wearable sensors, AI-assisted tools, and telerehabilitation to support independence, increase safety, enhance mobility, and improve the overall quality of life of older adults. The chapter also discusses the challenges associated with implementing technology-oriented rehabilitation. It outlines future directions to deliver personalised, accessible, and evidence-based rehabilitation that promotes independence and healthy ageing in older adults.

Keywords Fall Prevention · Functional Independence · Geriatric Rehabilitation · Older Adults

5.1 Physical Disability, Frailty, and Function in Older Adults

There has been a rapid increase in the global population's life expectancy. As of 2025, adults aged 60 and above account for 10% of the global population of eight billion (Cidre, 2025). According to the World Health Organisation (WHO) 2025 statistics, one in every six people will be 60 years or above by 2030. The number of people aged 60 and older is projected to increase from 1.1 billion in 2023 to 1.4 billion by 2030. It

S. Sheraz (✉)
Faculty of Rehabilitation and Allied Health Sciences, Riphah International University, Islamabad, Pakistan
e-mail: suman.sheraz@riphah.edu.pk

 95
A. N. Malik et al., *Emerging Technologies in the Rehabilitation of Physical Disabilities*, SpringerBriefs in Modern Perspectives on Disability Research, https://doi.org/10.1007/978-981-92-1343-6_5

is estimated that this increase will be doubled (2.1 billion) by 2050 and tripled (426 million) for people aged 80 years and above by 2050 (World Health Organization, 2025). The process of ageing is associated with a decline across multiple physiological systems of the body. This leads to a gradual decline in function and increased vulnerability to physical disability in older adults. Ageing is a multidimensional process affecting multiple systems of the body. The changes in these systems lead to frailty, reduce resilience to physical and psychological stressorsz, and impair the ability to perform functional tasks and Activities of Daily Living (ADLs) (Tenchov et al., 2023).

5.1.1 Age-Related Physiological Changes:

Ageing is a complex biological process that affects all the body's systems. With advancing age, physiological reserves decline and functional capacity decreases. This predisposes older adults to diseases and disabilities. Impairments across multiple body systems reduce strength, endurance, mobility, balance, and independence in older adults. With increasing age, muscle strength decreases, there is a loss of bone mass, increased fragility, and susceptibility to fractures (Tenchov et al., 2023). Ageing also affects neuronal structure and function, leading to reduced motor control, decreased reflexes, and impaired coordination. Cardiac output is also reduced, cardiovascular endurance is decreased, and, therefore, functional capacity is reduced. Moreover, their executive function, attention, processing speed, and ability to function independently also decrease. Also, the immune system deteriorates, increasing susceptibility to infections (Marzola et al., 2023). Sleep, vision, hearing, and proprioception gradually deteriorate, affecting balance and environmental awareness. As a result of these impairments, there is a gradual loss of physical fitness, leading to functional decline and decreased performance. Physical disability is a consequence of ageing, characterised by limitations in mobility, balance, strength, endurance, and the ability to perform ADLs. The synergistic effect of these multisystem impairments leads to mobility limitations, increases the risk of falls, and increases dependence.

5.1.2 Primary Impairments Leading to Physical Disability

Age-related physical disability is a consequence of three interconnected physiological systems that lead to progressive decline in the physical function of older adults. Musculoskeletal impairments resulting from reduced strength and mobility, sensory impairments compromising environmental awareness and spatial orientation, and cognitive deficits affecting motor planning and execution all result in functional limitations. These components not only contribute to declines in physical function independently but also exhibit multiplicative effects when combined. The older adult with muscle wasting might rely on vision for balance, but if vision is also compromised, the risk of falls is accelerated. Cognitive deficits further compromise anticipatory control and adaptive gait responses.

5.1.3 Musculoskeletal Deterioration: Sarcopenia and Mobility Limitation

Sarcopenia is defined as age-related progressive loss of muscle mass and strength. It is a fundamental contributor to physical disability in older adults, beginning as early as the third decade in men and the fourth decade in women. Sarcopenia causes a gradual reduction in the number and size of Type II muscle fibres, followed by infiltration of intramuscular fat. Multiple other factors accelerate muscle weakness resulting from sarcopenia. These include physical inactivity, nutritional deficits, and chronic health conditions that directly impact mobility and function. Also, bone microarchitecture weakens due to decreased bone mineral density (osteopenia and osteoporosis). This occurs particularly in females, increasing the risk of fragility fractures. Other musculoskeletal changes include cartilage deterioration, increased joint stiffness, pain, and reduced range of motion in weight-bearing joints (Gustafsson & Ulfhake, 2024; Li et al., 2021). Older adults are also at increased risk of injury due to reduced tendon and ligament elasticity. These changes in the musculoskeletal system led to functional limitations, including reduced strength, power, endurance, and gait speed. Chronic pain and stiffness also lead to activity avoidance and further deconditioning. This thereby increases the likelihood of falls and fractures, dependency in self-care activities, and an increase in the rate of hospitalisation.

5.1.4 Sensory Impairments Affecting Balance and Mobility:

Ageing significantly affects the sensory system by declining visual, proprioceptive, and vestibular function, which are responsible for maintaining balance and postural stability. This, in turn, becomes the most significant contributor to fall risk in older people. Hearing loss occurs in 25% of adults over 60 years of age (Hong et al., 2024). There is a loss of vestibular hair cells and impaired otolith function, leading to difficulty maintaining orientation during head movements. Also, visual impairment affects 20–22% of adults aged 70 and older. Macular degeneration reduces retinal nerve fibres and ganglion cells, and axonal atrophy occurs with increasing age. Increasing age also reduces joint sense, decreasing the number and sensitivity of proprioceptive receptors in muscles, tendons, and joints (Wang et al., 2024). Together, these sensory deficits affect balance and postural stability and increase dependence on compensatory strategies, increasing the risk of falls. This sensory decline is clinically manifested as unsteady gait, difficulty with transfers, and increased dependence on assistive devices.

5.1.5 Cognitive Changes Influencing Motor Performance

Cognitive deficits in older adults manifest as slowed walking and processing speed, diminished executive function, reduced attention span, and impaired dual-task performance. Slow cognitive processing speed plays a central role in the development of physical disability. The information processing is delayed, limiting the ability to integrate sensory input and generate a timely motor response. It is closely linked to mobility, balance, and postural control, with older adults exhibiting variable gait patterns. There is a delayed response to balance perturbations and decreased adaptability to environmental challenges. This reduces older people's confidence due to episodes of instability and near-falls. These effects are evident even with mild cognitive impairment because of the combined effects of cognitive, sensory, and motor decline, increasing the risk of falls and functional dependence. This reduction in physical activity leads to deconditioning and functional decline (Ma & Chan, 2020).

5.1.6 Frailty and Multimorbidity

Frailty is a state of reduced physiological reserve and increased vulnerability to various health outcomes. It is reported to occur in 51% of community-dwelling older adults aged 90 years and older. This geriatric syndrome results from deficits in muscular, endocrine, and neurological function. The core features include unintentional weight loss, exhaustion, weakness, slowness, and low physical activity. It ultimately leads to reduced muscle strength, increased fatigue, malnutrition, reduced exercise tolerance, and decreased social engagement (Yang et al., 2024). Frail older adults have an increased risk of falls, depression, disability, and dependency. Age-related decline in physical, psychological and social factors results in geriatric syndromes that act as a pathway to physical disability. Moreover, multimorbidity in older adults results in polypharmacy, including medications such as antihypertensives, sedatives, and psychotropics. These medications can impair balance, cognition, blood pressure and reaction time in geriatric patients. Drug interactions further exacerbate the impairments, resulting in functional decline.

5.1.7 Impact on Mobility and Activities of Daily Living

The cumulative effects of age-related multisystem impairments, sarcopenia, sensory deficits, cognitive decline, frailty, and geriatric syndromes are clinically presented as progressive functional limitations. Loss of IADL independence often represents the first major functional threshold toward dependency. These conditions compromise the functional capacity that leads to physical disability, reduced participation, and diminished quality of life. This ultimately increases the socioeconomic burden worldwide and underscores the need for rehabilitative approaches to prevent and manage functional limitations in older people.

5.2 Geriatric Rehabilitation: Overview, Limitations, and the Emerging Role of Technology

Geriatric rehabilitation is an intervention aimed at helping older adults with disabilities regain physical, psychological, and social skills to become more independent. The rehabilitation needs are growing for older adults, with the increase in average age, contributing to disability globally (Preitschopf et al., 2023). Healthy ageing is endorsed by the World Health Organisation (WHO), with a focus on promoting functional ability and reducing disability among older adults (World Health Organization, 2023). Geriatric rehabilitation focuses primarily on promoting healthy ageing and improving the quality of life of older adults.

5.2.1 Overview of Geriatric Rehabilitation

Geriatric rehabilitation involves a tailored approach that addresses the comprehensive rehabilitation needs of older adults. This is important to address impairments such as sarcopenia, frailty, and balance deficits, each of which is contributing to functional decline and increased dependence in ADLs. Geriatric rehabilitation targets key geriatric syndromes, aiming to restore functional independence and mobility in older adults. This is achieved through various interventions, including exercise therapy, adaptive equipment, assistive devices, and functional task modification, with an emphasis on patient-centred goals (Lim, 2025; Preitschopf et al., 2023; Wong et al., 2024).

Geriatric rehabilitation programs can be delivered as inpatient, geriatric day hospital, outpatient, nursing home, and community-based services. According to the meta-analysis, outpatient geriatric rehabilitation is as effective as usual care in improving functional performance, reducing length of hospital stays, lowering re-admission rates, and improving quality of life for both patients and caregivers (Preitschopf et al., 2023). According to another systematic review & meta-analysis, geriatric rehabilitation in in-patient or day hospital setting effectively reduces rate of mortality (risk ratio (RR 0.84, 95% confidence interval (CI) 0.76–0.93), long-term care home admissions (RR 0.86, 95% CI 0.75–0.98), and improves functional status (Standardised Mean Difference (SMD) 0.09, 95% CI 0.02 to 0.16), and cognition (Mean difference of mini-mental status exam score 0.97, 95% CI 0.35 to 1.60). Geriatric rehabilitation follows a multifaceted approach addressing key impairments such as frailty, functional decline, impaired gait, mobility limitation, balance deficits, risk of fall, and cognitive decline. To address these impairments, multicomponent programs including exercise training with nutritional support, cognitive training, and patient education about the disease are implemented. Alongside this, the geriatric rehabilitation team also focuses on multimorbidity, polypharmacy, and geriatric syndromes. Patients' progress is regularly monitored, and rehabilitation goals are reviewed accordingly. The goal of the entire team is to improve the quality of life, enhance social participation, and promote independent living for older adults (Lim, 2025).

While the team focuses on a range of interventions to improve patient outcomes, including medical management, occupational therapy, cognitive training, and social support, physical therapy plays a key role in restoring function, preventing falls and fractures, improving mobility, and preventing physical disability in older adults. Exercise is considered a first-line treatment for frailty, sarcopenia, and functional decline. Customised exercise plans that incorporate aerobic exercises, strengthening, balance, gait training, and functional task performance training are key evidence-based conventional physical therapy approaches (Izquierdo et al., 2025). A recent systematic review and meta-analysis recommended multicomponent training, with 170 min (approximately 3 h) of weekly exercise spread across 1 to 7 sessions, to improve physical function. This should include strength, endurance, and balance training, with the intensity adjusted as the patient improves to keep the training load constant; the programme must be maintained long-term to prevent benefits from disappearing (Valenzuela et al., 2023).

5.2.2 Limitations of Conventional Rehabilitation Protocols

Conventional geriatric rehabilitation has demonstrated clinical utility for older adults; however, certain challenges and limitations remain. These include limited continuous monitoring, delayed detection of declines in patients' current health conditions, such as the risk of falls or cognitive impairment, and safety risks at home. One of the core limitations is that supervision is confined to the scheduled clinical encounters. If the supervised sessions are scheduled weekly or biweekly, it becomes difficult to ensure adherence and exercise quality owing to poor cognition, multimorbidity, and fluctuating health status. Moreover, changes in functional status, fall risk, and cognitive status are identified only during scheduled follow-up sessions or during an adverse event between scheduled visits. Subtle declines in mobility, balance and activity levels go unnoticed for weeks or months. This delayed detection can have severe consequences, given the rapid functional decline caused by deconditioning or disease exacerbations. Standard clinical assessment tools, such as timed walk tests and balance assessments, provide a snapshot but fail to capture the day-to-day decline in functional impairments. Moreover, conventional treatment approaches demand time from the patient, caregiver, and therapist, as well as their availability, adherence to the treatment, and frequent hospital visits for supervised sessions (Dore et al., 2023). During unmonitored activities, safety risks can occur at home, including falls, medication errors, and adverse reactions like dizziness or hypoglycemia. Limited objective feedback, inability to sustain patient motivation and engagement, and inadequacy are also the limitations of conventional rehabilitation protocols. Another limitation in low- and middle-income Countries (LMICs) is the infrastructure deficit and privacy concerns that hinder the adoption of conventional geriatric rehabilitation.

5.2.3 *Role of Technology in Geriatric Rehabilitation*

Emerging technologies offer solutions to the challenges and limitations of conventional geriatric rehabilitation. These technologies enable continuous monitoring and recording of data on older adults' functional status. This monitoring can be done between scheduled supervised sessions in hospital settings. Wearable sensors and monitoring systems can track improvements or declines in patients' performance over time. Sensors and mobile devices can record movement patterns, activity levels, and functional outcomes such as gait speed and balance. Wearable devices, along with accelerometers, Inertial Measurement Units (IMUs), or position-tracking, measure and track exercise performance (Zhang et al., 2025). Recorded data can detect deviations from defined patterns and help detect deterioration in health parameters early, enabling early prevention of complications. Robotic systems help improve patient repeatability and adaptability, and enhance movement precision. It not only provides motor feedback but also sensory feedback, including visual, auditory, and haptic, to students. Robotic technologies can adjust and optimise exercise therapy for patients by providing precise, real-time feedback on their performance and progress. Integrating virtual reality-based gaming into the exercise session has made it more engaging and motivating for patients. Technological developments such as Virtual Reality (VR) and Augmented Reality (AR) provide interactive environments that enhance patient motivation and adherence (Dore et al., 2023). They provide real-time feedback and a safe, controlled environment, increasing confidence and reducing the fear of falls among older adults. Instead of using generalised exercise protocols for all patients, technology can provide customised exercise plans based on patients' performance and feedback. AI-driven platforms analyse patients' data and perform predictive modelling to customise exercise plans. These systems can adjust exercise intensity, frequency, and type based on functional performance data. Smart home sensor technologies detect environmental changes, detecting fall risk, prolonged inactivity, or abnormal movement patterns. They can be categorised into systems that monitor physiological, functional, safety, security, social interaction, and, lastly, cognitive and sensory assistance (Tian et al., 2024). They automatically alert caregivers or clinicians, allowing continuous tracking of mobility, activity levels, gait patterns, and transfers. This is particularly important for older adults with balance deficits, sensory impairments, or cognitive decline, as it reduces the risk of unattended adverse events.

5.3   Emerging Technologies in Geriatric Rehabilitation

Emerging technologies show promising results in improving mobility, functional independence, and overall quality of life of older adults. These digital health interventions include robotics and assistive devices, virtual reality, augmented reality, exergaming, wearable technologies, artificial intelligence (AI) applications, telerehabilitation, smart home systems, and Ambient Assisted Living (AAL). Digital health technologies can support real-time tracking and monitoring of health parameters. They can help with early risk detection, personalised exercise and cognitive training programs, and the adoption of healthy behaviours, contributing to improved physical and cognitive functioning. Digital interventions can empower older adults to engage proactively in their own health management rather than relying solely on in-person care. The details on these technological interventions, their efficacy, usage, and protocols are explained below:

5.3.1   Robotics and Assistive Technologies

Robotic systems integrate motor, sensor, and control systems to execute precise, repetitive therapeutic movements (Banyai & Brişan, 2024). Assistive robotic systems are being designed to support older adults with mobility, hygiene, and even household chores. Various sensors and actuators are integrated into them to assist older adults with their physical tasks. Robot-assisted rehabilitation sessions enable more efficient training sessions with increased therapy dosage. Rigid exoskeletons provide mechanical support and help improve strength and postural control. However, they have limitations, such as safety concerns, ease of use, and cost. Soft exoskeletons, on the other hand, are flexible and lightweight and are used by older people to assist with ADLs. They utilise actuators, body attachments, sensing systems, and control algorithms, and address the limitations of rigid exoskeletons (Lingampally et al., 2024). Robots for ankle rehabilitation operate both as exoskeletons that improve gait patterns and as platform robots for exercise and ankle motion. Robots for upper-limb rehabilitation act similarly to end-effector robots for older adults with upper-limb dysfunction. Sports Rehabilitation Robots (SRRs) provide motor rehabilitation for older adults, specifically those with physical disabilities resulting from degenerative diseases. These SRRs are widely developed and classified as exoskeletons (most widely used), end-effector robots, smart walkers, and intelligent robot rollators (Ju et al., 2023; Lingampally et al., 2024).

A systematic review of the impact of care robots on older adults concluded that they could assist with a range of tasks, from physical tasks to emotional support and overall health management (Lee & Yu, 2025). These care robots include doll-type robots that provide emotional support through sensory interaction. Robot-type care robots have a human-like appearance and assist with physical activities such as exercise. AI integrated care robots are tablet-type robots, and they help in health management through health alerts, e.g. for fall reduction, and therefore reduce hospital stay. There are also combined-type robots that merge two or more primary functions to address the needs of older people. Another category of robots is Socially Assistive Robots (SARs) that can help older adults through monitoring, assessment of cognitive and physical changes and assistance with ADLs. SARs can therefore reduce caregiver burden by integrating multiple care functions into a single system tailored to individual users' needs (Robinson & Nejat, 2022). They can provide companionship to older adults and help them maintain their health by monitoring vital signs, giving reminders, assisting with tasks, and signalling when the patient is at risk of falling (Maranesi et al., 2022). However, there is a need to design a SAR that accounts for older adults' limited familiarity with technology and ensures positive human–robot interaction. Wearable robotic exoskeletons, including rigid exoskeletons and soft exosuits, help in gait rehabilitation and mobility assistance in both clinical patients and healthy older adults with age-related changes and functional decline. Rigid exoskeletons are better suited for severe gait impairments due to their strong joint support, whereas soft exosuits provide lightweight assistance for milder deficits (Cha et al., 2025). Robotic hip exoskeletons are also a variant that provides support for older adults and rehabilitation for patients with gait impairment. They enhance performance during normal walking, loaded walking, and manual handling of heavy tasks through assistive torque (Cha et al., 2025).

Overall, robotic rehabilitation is a valuable addition to conventional rehabilitation approaches for older adults. Through its multiple benefits, it has proven to be an efficient emerging technology that addresses the needs of the ageing population.

5.3.2 *Virtual Reality, Augmented Reality, and Exergaming Technologies*

Virtual technologies offer benefits across physical, cognitive and psychological domains and can be used for the assessment and intervention of community-dwelling older adults. According to Benjamin's systematic review, VR has been shown to improve balance, reduce fear of falling, and correct gait patterns in older adults (Robinson & Nejat, 2022). VR is a transformative technology that motivates and engages older people through gamification and real-time feedback in a safe and controlled environment. A recent systematic review reported improvement in balance, postural control, and functional mobility through immersive VR interventions. However, further research is needed to determine the standardised protocol for each outcome and the long-term effectiveness of VR-based interventions (Santana Muñoz et al., 2024).

VR interventions have been shown to improve physical and psychological well-being by enhancing cognitive function and reducing the risk of falls among older adults(Aldardour & Alnammaneh, 2025). It also helps improve gait, postural control, and dual-task performance in older adults (Corregidor-Sánchez et al., 2021). A meta-analysis and systematic review found that more than 18 VR training sessions improve functional mobility in older adults. Another systematic review has demonstrated the effects of VR interventions on gait, fall prevention, and static and dynamic balance compared with conventional balance training in healthy older adults (Rodriguez-Almagro et al., 2024). It also reported 30% greater treatment adherence in the VR group than in the conventional exercise therapy group. Enjoyment, improved motivation, and increased adherence lead to more successful completion of exercise sessions, thereby resulting in long-term functional improvement. The exercise protocol ranged from 60-min sessions for 6 weeks to 30-min sessions for 8 weeks. The outcomes that improved in older adults included increased muscle strength, faster reaction time, reduced body sway, improved gait parameters, alertness, hip range of motion, leg-shoulder synchrony, mobility, decreased dizziness, and improved postural control. A meta-analysis reported lower VR sickness with non-immersive VR than immersive VR (Høeg et al., 2021; Rodriguez-Almagro et al., 2024).

A systematic review reporting the significant effects of AR-based interventions for lower-limb rehabilitation cited seven studies that examined healthy older adults, focusing on outcomes such as balance, gait, muscle strength, physical performance, and fall efficacy (Chang et al., 2022). Another scoping review reported that wearable AR systems improve physical health, cognitive support, ADLs, emotional, and experiential engagement (Mikhailova, 2025). A recent systematic review (2025) reported the use of AR for cognitive support, physical activity guidance, navigation, and social engagement. Each study in the review had different devices, such as smart glasses, tablet-based AR, or wearable AR. Session frequency and duration were not consistently reported. Some studies involved a single exposure to AR, while others allowed repeated short sessions lasting 10 to 30 min (Ceyssens et al., 2025).

VR and AR technologies offer innovative approaches to enhance balance, mobility, cognitive function, and engagement in older adults. Evidence suggests that immersive, task-oriented, and feedback-driven VR interventions can improve gait parameters, postural control, and functional mobility. Combining these technologies with wearables, smart environments, and telerehabilitation platforms can extend access, increase adherence, and optimise outcomes in geriatric rehabilitation.

5.3.3 Wearable and Sensor-Based Technologies

Wearable technologies are electronic devices that monitor and record physical activity and health-related outcomes. Wang et al. have categorised the WDs for monitoring older adults into indoor positioning systems to track location within indoor environments, real-time physiological monitoring devices that focus on continuous measurement of vital signs, and activity recognition systems to identify and quantify daily activities, posture, gait, and functional mobility (Wang et al., 2017).

The most used Wearable Devices (WDs) are accelerometers, which provide information on step count, sedentary behaviour, energy expenditure, and time spent on physical activities of varying intensities. These parameters affect glycemic index, and tracking them can support diabetes management. Recent studies have reported that accelerometer-assessed high-impact activities are positively associated with bone health, whereas sedentary behaviours, including prolonged sitting, are negatively associated with it. These accelerometers can also provide users with daily feedback on whether they achieve sufficient mechanical loading to protect against age- and menopause-related bone loss (Teixeira et al., 2021). Wrist-worn WDs that monitor heart rate use optical sensors and commonly employ photoplethysmography to non-invasively measure heart rate at low cost and over longer periods. WDs estimate Blood Pressure (BP) indirectly, using algorithms derived from pulse wave velocity, pulse transit time, or photoplethysmography signals. These approaches help with frequent, continuous BP estimation, but their accuracy remains inconsistent, particularly for older adults. Teixeira et al. reported that WDs for BP estimation can be used for trend monitoring but cannot replace conventional BP measurement in older adults because age-related physiological changes, such as arterial stiffness, reduced vascular compliance, and endothelial dysfunction, significantly affect pulse wave characteristics. Other devices that monitor blood pressure include wristwatches, in-ear devices, and glasses equipped with optical-based pulse sensors (Olmedo-Aguirre et al., 2022; Teixeira et al., 2021). A headband worn overnight can detect sleep and wake patterns and biomarkers, using sensors to measure electroencephalography (EEG), electrocardiography (ECG), pulse rate, head movement, and the number of snoring incidents (Olmedo-Aguirre et al., 2022). There are intradermal sensors for real-time monitoring of blood glucose levels. It records readings every 5 min and then sends them to the system for 6 days, after which they should be removed or discarded. Some smartwatches and bracelets include an oximeter, optical heart rate sensors, peripheral temperature sensors, a 3-axis accelerometer, a gyroscope, an ECG, and a barometric altimeter for monitoring relevant physiological parameters. The most common types of WDs used in older adults are watches, bracelets, patches, intradermal sensors, and portable sensors (Olmedo-Aguirre et al., 2022).

A qualitative systematic review and meta-synthesis that synthesised the experiences of older adults using WDs compiled views from 349 participants aged 51–94 years across 20 qualitative studies. Older adults perceived WDs as beneficial tools for monitoring health-related parameters, particularly for tracking physical activity, vital signs, and daily routines. However, the major challenge was that devices were often described as uncomfortable, bulky, or difficult to operate. Another challenge was the complex interfaces, small displays, and frequent charging requirements, which affected long-term adherence to the WDs. This meta-synthesis highlighted that older adults were more likely to engage with wearables when the data were easy to understand, linked to personal health goals, and supported by their clinicians or caregivers. They had concerns related to privacy, data security, and autonomy. Some participants expressed discomfort with continuous monitoring, fearing loss of independence, particularly when data were shared with family members. Support from healthcare professionals, family encouragement, and proper training facilitated acceptance, whereas a lack of guidance and technical support led to discontinuation of device use. WDs are now also used for frailty screening and early detection, as well as for objective assessment through wearable inertial sensors. Assessment of gait, fall risk, upper-extremity functional tests, sit-to-stand test, and tracking and measurement of physical activity in clinical and home-based settings can help with early assessment and treatment of frailty (Yixiao et al., 2025).

Sensors connected through the Internet of Things (IoT) can enable continuous, real-time monitoring, helping patients manage dementia, Alzheimer's disease, Parkinson's disease, cardiovascular disease, and Frailty, and promoting healthy, active ageing. Integration with cloud-based platforms enables remote access to data for caregivers and healthcare professionals, facilitating timely intervention (Stavropoulos et al., 2020). A systematic review reported that wearable sensor-based devices are an effective approach to assessing gait in people living with dementia. The studies included in the review primarily employed Inertial Measurement Units (IMUs), accelerometers, and gyroscopes, commonly placed on the lower back, waist, or lower limbs, to capture gait parameters during walking tasks in both laboratory and real-world settings. According to the review, wearable sensors can reliably detect gait abnormalities associated with dementia, including reduced gait speed, shorter step and stride length, increased gait variability, impaired symmetry, and altered cadence. These gait changes were often more pronounced during dual-task walking (Weizman et al., 2021).

In short, WDs have potential applications in cardiac rehabilitation, diabetes management, fall prevention, and frailty monitoring. Most of these devices have not been validated across multiple populations and are not designed specifically for older adults, leading to usability issues.

5.3.4 *Artificial Intelligence and Data-Driven Rehabilitation Systems*

Artificial intelligence (AI) involves multidisciplinary knowledge and techniques to simulate and extend human intelligence through machine learning, data-driven algorithms, robotics, decision support systems, and pattern recognition (Wang et al., 2023). Studies on the role of AI in older adults' health care reveal the use of AI technologies integrated with rehabilitation, including robots, exoskeletons, wearable sensors, intelligent homes, and smartphone applications (Ma et al., 2023). A scoping review identified five roles of AI-based technologies in addressing the unmet healthcare needs of older adults. These include the roles of rehabilitation therapist, emotional supporter, social-interaction helper, supervisor, and cognitive trainer (Ma et al., 2023). According to a recent meta-synthesis review, AI-based systems significantly improve geriatric care management, thereby enhancing quality of life. These systems use machine-learning-based algorithms to predict health outcomes, enable early disease diagnosis, and design customised plans for disease prevention and chronic disease management. AI integration into sensors and smart apps can help with home-based patient monitoring and enable effective communication between older adults and their healthcare providers (Sharafi & Mohammadyari, 2025).

A systematic review by Leghissa et al. evaluated 41 studies that applied Machine Learning (ML) to detect, predict, or classify frailty in older adults. Most studies used conventional ML models (e.g., Support Vector Machines, Random Forests, Logistic Regression, and Neural Networks). However, only one study developed a novel ML framework specifically for frailty classification. Frailty-related data were obtained from sensor-based data (e.g., gait data collected via wearable sensors (accelerometers/IMUs)) and from medical or clinical records, including demographic, cognitive, functional, and biomedical variables (Leghissa et al., 2023). Another review by Ayushi Das evaluated 70 papers on the increasing use of ML in diagnosing, treating, and managing geriatric diseases. It also highlighted the potential role of ML in improving the quality of life of the ageing population and generating indices such as the successful ageing index. Among the neurodegenerative diseases, the most common disease was Alzheimer's disease, followed by dementia and cognitive dysfunction. Prediction models based on supervised ML have been developed to predict disease and identify risk factors for non-communicable diseases such as diabetes, hypertension, chronic renal disease, cancer, and cardiovascular diseases. These ML algorithms enable timely interventions for disease prevention or progression by early identification of patients at high risk. However, validation of these models across diverse demographic groups is still needed to improve geriatric care (Das & Dhillon, 2023).

AI-driven Virtual Physiotherapy Assistants (VPA), combined with wearable sensors, are another modality that provides real-time feedback and customised support for home-based rehabilitation. These AI-powered systems use wearable sensors, such as motion trackers, smart bands, and pressure sensors, to collect data on patients' physical movement, posture, range of motion, and speed. This data is analysed by AI-driven systems that detect errors in movement patterns, followed by corrective feedback to the patients. This mitigates the risk of injury associated with incorrect home exercise techniques. Few systems have features that notify the physiotherapist when patients exhibit consistent errors in exercise performance. The challenges encountered with these systems include sensor accuracy, security and privacy concerns, and the costs of wearable sensors, especially in low-income regions or areas with limited access to advanced healthcare technology (Olawade et al., 2025).

Another systematic review identified the different AI-based applications used for the physical rehabilitation of individuals with physical disabilities, along with the barriers and enablers of their use in health care. These applications help improve outcomes by initially interpreting large volumes of complex data, identifying patterns, and tailoring interventions to patients' needs. The studies included in the review explained that AI can help patients restore their function in different ways. A study used AI to analyse and interpret motion data related to gait issues. Another study used an app-based health promotion system to conduct predictive analysis in patients with back and neck pain. Other studies included in the review used wearable sensors, smart watches and apps, Brain–Computer Interface (BCI),

Functional Electrical Stimulation (FES), analysed motion data using a wearable inertial system, improved upper-limb motor function using VR systems, and used a variety of AI-based mobile applications. Another systematic review highlighted the role of AI-powered social robots in promoting physical activity among older adults. These AI-integrated social robots motivate older adults to engage in physical activities. This, in turn, led to longer training duration, improved exercise adherence, and higher engagement among older adults. The studies in the review revealed that social robots have proven effective in both structured environments, such as rehabilitation centres and nursing homes, and in older adults care facilities and community centres (Sumner et al., 2023).

5.3.5 Telerehabilitation and Digital Care Delivery Platforms

Another emerging technology in geriatric rehabilitation to overcome economic and geographic barriers is telerehabilitation. Telerehabilitation is the provision of home-based rehabilitation services under the guided supervision of a healthcare professional via communication technologies. A growing body of knowledge supports the effectiveness of telerehabilitation for older adults with musculoskeletal conditions, chronic disease management, and postsurgical patients, promoting independence, functional performance, physical function, and quality of life. Telerehabilitation and digital health interventions tend to promote healthy ageing in older adults. Its adoption becomes easier among the older adults with higher education and greater digital literacy. However, low confidence, cognitive or sensory impairments, and fear of technology make their engagement in digital interventions challenging. These digital interventions have proven beneficial for physical, mental, cognitive, and preventive health. Long-term engagement improves when the technology is supportive, with simple interfaces, large font size, and clear instructions (Alruwaili et al., 2023).

A systematic review based on 26 Randomised Controlled Trials (RCTs) has shown that telerehabilitation is a promising alternative to in-person rehabilitation for improving functional performance in community-dwelling older adults. Another systematic review and meta-analysis assessed the effects of telehealth-delivered exercise interventions on physical functioning in older adults with frailty, cognitive, or mobility disability. In the included studies, telehealth was delivered through both synchronous (real-time interaction between the participant and the healthcare provider) and asynchronous (sharing of educational materials/videos) modes. Pooled analyses demonstrated moderate effect sizes favouring telehealth exercise for mobility and muscle strength, and a small effect for balance (Gamble et al., 2024). Telehealth exercise interventions were found to be feasible with good acceptance and adherence, and minimal adverse events were reported (Dawson et al., 2024). A scoping review by Kebede et al. maps the literature on older adults' engagement with digital technology. He has defined a three-stage continuum for conceptualising digital engagement. These stages are nonuse, where individuals do not adopt technology. A stage of initial adoption where decisions are made to accept or reject digital tools, and sustained engagement where there is continued and meaningful use of digital technologies over time (Kebede et al., 2022). The review included 96 studies, most of which focused on the initial adoption phase, and a few investigated nonuse or sustained engagement, indicating a gap in the literature. Everyday technologies (like smartphones, the internet, and computers) were most frequently studied (Kebede et al., 2022).

Recent technologies have also incorporated Remote Monitoring Systems (RMS) for high-risk patients with a history of hospitalisation or with chronic co-morbid conditions. RMS comprises telemonitoring devices and sensors that detect blood pressure, oxygen saturation, etc. They may also include health questionnaires for patients or their caregivers to complete. Automated alerts and dashboards that detect abnormal readings and notify the response team are also part of it. The team then reviews the data and takes necessary action. As per the results of a scoping review, the RMSs improved the quality of life and reduced unplanned hospital visits by timely detection of abnormal readings (Salma et al., 2025). A meta-analysis by Wick et al. reported improved physical function outcomes, including mobility, strength, balance, and functional performance, with a physiotherapist-led, exercise-based telerehabilitation program. It also improved patient outcomes, including quality of life, greater confidence, fewer hospital admissions, and fewer emergency visits (Wicks et al., 2023).

5.3.6 Smart Home Systems, Ambient Assisted Living (AAL), and Internet of Things (IoT)

Smart Home Systems, Ambient Assisted Living (AAL), and the Internet of Things (IoT) are interconnected technological innovations designed to enhance independence, safety, and quality of life for individuals with physical disabilities. The IoT forms the basis of these technologies and refers to a system of internet-connected physical devices. These devices are embedded with software, sensors, and communication technologies that enable them to collect data, process large volumes of data over a cloud network, and make AI-supported decisions. IoT-enabled devices include wearable activity trackers, physiological monitors, environmental sensors, and assistive mobility technologies. These systems allow continuous monitoring of remote patients' health parameters and support data-based clinical decision-making.

Smart home systems, devices, and environments provide safety, accessibility, and convenience to the patients. These systems integrate motion sensors, automated lighting, voice assistants, and security technologies that can be controlled via Wi-Fi or smartphone applications. Smart home systems not only monitor data but also feature technology that enables devices to communicate, automate tasks, and be controlled remotely. These systems can enhance patients' comfort and security by adapting to environmental factors such as temperature and light. Smart homes, therefore, support functional independence, improve safety (e.g., by reducing fall risk), and enable remote supervision when necessary (Shah et al., 2025). A systematic review by Ghafurian et al. covered 48 studies conducted in participants' homes, research labs, care centres, and hospital environments. The studies used different sensors with older adults. The sensors tested mostly were motion and infrared sensors, contact sensors, smart lighting, and automated lights. Few studies have used ambient smart devices, such as voice-controlled assistants, smart outlets, and automated switches, to control household functions. The studies assessed sensor accuracy, device responsiveness, and user acceptance, satisfaction, and ease of use. According to this review, smart home technologies are technically feasible and accepted by older users in both controlled and natural settings. However, the evidence demonstrating health or lifestyle benefits remains limited and in early stages (Ghafurian et al., 2023).

Ambient Assisted Living (AAL) refers to a health-oriented application of smart home and sensor technologies designed specifically for older adults, people with disabilities, and those with chronic conditions. Its primary goals are to promote independence, detect risk of fall or inactivity, and support ageing in place. It includes fall-detection systems, medication reminders, and emergency-response systems (Guerra et al., 2023). A recent systematic literature review highlighted the integration of Machine Learning (ML) algorithms, transforming these systems from simple monitoring tools into predictive, personalised healthcare platforms. Within AAL, Human Activity Recognition (HAR) is a system that automatically detects and classifies a person's activities using sensor data. This HAR works by collecting raw sensor signals and then cleaning noise, thereby normalising data. It then identifies relevant patterns (e.g., movement speed, orientation) and classifies the data using algorithms such as machine learning or deep learning. There is continuous ongoing development working to make AAL systems more effective in supporting independent living and preventing functional decline among older adults.

IoT-based smart home systems can improve quality of life, provide timely alerts during emergencies, and reduce dependency on caregivers. They provide services in four primary areas, including healthcare and nutrition, environmental control, safety and emergency management, and social engagement (Yamout et al., 2023). These IoT-based smart home systems form a foundational element of AAL, creating safe, responsive, and personalised environments that support independent living for older adults.

5.3.7 Non-invasive Brain Stimulation: Transcranial Direct Current Stimulation (tDCS)

Transcranial Direct Current Stimulation (tDCS) is a non-invasive brain stimulation technique that delivers a weak, constant electrical current (usually 1–2 mA) through electrodes placed on the scalp. This modulates neuronal excitability, improves cognitive function, true recognition, and working memory, and reduces false memories in older adults. This is a safe method to reduce the risk of falling, improve balance, and overall cognition in older adults. According to a systematic review, tDCS has shown good short-term effects on balance and gait patterns, with procedures considered completely safe and without adverse effects. The intensity from 1 to 2 mA, 20 min, 2–6 sessions over the cerebellum is an effective protocol for reducing the risk of falls in older adults (Bueno et al., 2024).

According to the results of a triple-blinded randomised controlled trial, the addition of tDCS to multiple-component training for functional capacity potentiated the effects of training on walking capacity, functional independence, balance, and quality of life at 30 days post-intervention (Correa et al., 2023). Anodal tDCS combined with unstable resistance training for eight weeks has also shown promising effects on balance, proprioception, and postural control, reducing the risk of falling in older adults. This integrated treatment approach, which simultaneously neuromodulates central-peripheral pathways, appears to be a promising fall-prevention strategy in geriatric rehabilitation (Peng et al., 2025).

5.4 Technology-Enhanced Geriatric Rehabilitation Protocols

The following sections outline technology-enhanced rehabilitation protocols targeting key functional domains of older adults, including falls, balance, gait, and functional mobility. These protocols, guided by systematic reviews and meta-analyses, apply technology to systematically improve balance, gait, and functional mobility in older adults.

5.4.1 Fall-Focused Digital Interventions

Age-related changes in balance and gait parameters lead to falls at least once a year in more than one-third of people aged 65 years and above (Dore et al., 2023). A meta-analysis compared different technology-based interventions as community-based fall prevention strategies. The interventions included telerehabilitation, exergaming, virtual reality, and mobile applications for balance training. The strongest evidence for fall reduction was observed with the telehealth-delivered exercise programs, followed by exergaming, which was also good for balance training. For fall detection and monitoring, smart home sensors were appropriate. Mobile-based apps were also helpful when used consistently, in combination with support and supervision, for older adults (Lee et al., 2024). VR can reduce fall risk in patients, similar to conventional exercises, but in half the time: 2 sessions per week, each lasting 30 min, for 6 weeks (Dore et al., 2023). The review reported a study by Phu et al. showing a small but significant decrease in the risk of falls in healthy older adults when assessed using the Falls Efficacy Scale (FES-I) (Dore et al., 2023). Another systematic review found that VR interventions were more effective than physical exercises in improving static and dynamic balance and reducing the risk of falls.

5.4.2 Technology-Based Interventions for Balance

VR training with 3D video games improved static balance by reducing medio-lateral and antero-posterior sway and the velocity of displacement. Dynamic balance was also improved, as evidenced by higher scores on the Berg Balance Scale, Functional Reach test, Timed Up and Go test, and lower-limb strength (Rodriguez-Almagro et al., 2024). Among the different balance training protocols, Nintendo Wii/Wii Fit Exergames included exercise sessions lasting approximately 30 to 40 min, with 2 to 3 sessions per week of balance and rhythm games lasting over 4 to 8 weeks. Kinect Game-Based Training was employed using full-body motion capture with visual feedback for 40–60 min per session. The sessions meet 3 times a week for 8 weeks. 3D Video Game Training with immersive 3D interactive tasks had 12 sessions, each 60 min. BioRescue Cognitive-Motor VR, focusing on dual-task balance challenges, was performed for 18–30 min, twice a week, for 6 weeks (Rodriguez-Almagro et al., 2024).

Meng Lieu reported in a meta-analysis that, for balance training, predominantly non-immersive VR (Nintendo Wii Fit™, Kinect-based systems) was used with community-dwelling older adults; however, some studies used immersive or semi-immersive VR with projected environments. Static balance tasks included single-leg stance, weight-shift control, and quiet standing with visual perturbations. However, dynamic balance tasks include multidirectional reaching, postural reactions to visual stimuli, and obstacle avoidance in virtual environments. Session duration for most interventions included 30 to 50 min, with 2 to 3 sessions per week. The effective cumulative dose exceeds 800 total minutes, and the intervention lasts 6 to 12 weeks. Significant improvements were seen in One-Leg Stance (OLS) and Postural stability measures. VR balance training showed greater effects than conventional balance exercises, especially when immersive elements and progressive difficulty were incorporated (Liu et al., 2022).

A scoping review that did a comparative analysis of different wearable technologies for balance rehabilitation reported that most interventions used inertial sensors, wearable accelerometers, smart insoles, or body-worn motion sensors to quantify postural sway, gait parameters, and dynamic balance during standing and walking tasks. Several studies included in the review reported significant improvements in clinical balance measures with wearable-assisted training, including reduced postural sway, improved gait stability, and better performance on functional balance tasks. Interventions incorporating real-time feedback were effective, as feedback helped participants correct posture, adjust weight shifting, and improve movement symmetry during exercises. Studies have demonstrated that wearable sensors provide more sensitive and objective detection of balance deficits than conventional clinical scales. WDs captured small changes in trunk motion, centre-of-mass displacement, and gait variability that are often missed during routine assessments, improving fall-risk identification and monitoring of rehabilitation progress (Nairn et al., 2025).

5.4.3 Gait Training Through Innovative Technologies

The older adults worked at about 4.7 times training intensity compared to conventional therapy, by taking around 1000 steps in a 30-min session with Robot-assisted Gait Training (RAGT) (Cha et al., 2025). Gait patterns can also be improved via VR interventions. By utilising the headset's embedded screen, the participant's virtual foot takes a longer step than in reality, exaggerating the reduced step length on the more affected side. This manipulation encourages the user to take more symmetrical steps, balancing stance and swing times between both legs. As a result, the cadence of the affected side becomes more regular, resulting in a more uniform overall gait pattern (Doré et al., 2023).

VR showed greater positive effects on gait parameters and functional gait scores than conventional approaches. Gait speed and spatial parameters: Several studies reported greater improvements in gait speed and spatial parameters, including stride length, walking speed, and step symmetry, with VR compared with treadmill or usual care. Improvements in functional gait assessment were noted in VR groups, indicating better gait stability. Gait training protocols included VR-based gait training vs treadmill alone, where training integrated VR feedback during gait tasks. In another protocol, virtual environments requiring dynamic weight shifts and stepping responses were linked to visual targets. The duration of the session in most protocols ranged from approximately 30 to 60 min. 2–3 sessions per week were conducted for 4–16 weeks across different studies (Rodríguez-Almagro et al., 2024).

For gait training, the meta-analysis by Meng Lieu reported that immersive VR with a treadmill, projected or head-mounted environments, and semi-immersive systems with visual flow and obstacle negotiation were effective. Core gait tasks included overground or treadmill walking with visual flow manipulation, step-length and cadence targets, virtual obstacles and pathways, and dual-task walking, with motor tasks combined with visual or cognitive tasks. Session duration varied from 30 to 40 min, with 2 to 3 sessions per week. The intervention lasted from 6 to 15 weeks and resulted in improvement in gait speed, step length, and symmetry. There were inconsistent findings in a few studies, but overall, immersive VR and longer intervention duration showed greater effects (Liu et al., 2022).

5.4.4 Technological Interventions for Functional Mobility

A meta-analysis by Meng Lieu reported that both immersive and non-immersive VR systems that emphasised task-oriented movement sequences rather than isolated exercises were effective for improving gait parameters. Core mobility tasks included Sit-to-stand transitions, turning and directional changes, stepping responses to visual cues, and combined standing, walking, and turning tasks. The session duration lasted from 30 to 45 min, with 2 to 5 sessions per week. The effective weekly dose was > 120 min per week, and more than 8 weeks of intervention showed better outcomes (Liu et al., 2022). According to a systematic review, functional performance in older adults is influenced by several factors, including lower-limb strength, endurance, balance, and postural control. A systematic review found that telerehabilitation improves functional performance among community-dwelling older adults. The performance tests used to determine functional performance were valid and reliable, including the Six-Minute Walk Test (6-MWT), Timed Up and Go test, Berg Balance Scale, Sit-to-Stand, and 10 m walk test. Intervention duration ranged from 3 weeks to 6 months, with 12-week protocols being the most common. Individual rehabilitation sessions typically lasted 20–60 min, with 1-h sessions most frequently reported. The frequency of supervised sessions ranged from twice daily to once weekly. The duration of the study intervention varied from 3 weeks to 6 months, with the majority being 12 weeks. Functional performance was improved through various exercises, including endurance, resistance, and balance training, and was performed individually and in combination. To address safety concerns, a few studies had participants perform exercises in the presence of a family member or caregiver. In a few studies, pre-intervention exercise testing was performed. In others, real-time monitoring was conducted remotely using heart rate monitors, SpO2 devices, and other equipment to monitor patients during exercise.

5.5 Challenges in Technology-Enabled Geriatric Rehabilitation

Despite technological advances in geriatric rehabilitation, certain challenges are associated with these technologies. Key challenges can be categorised into patient-related, technological, financial, infrastructure, and workforce-related factors. Lack of technology, data privacy, internet connectivity issues, limited digital literacy, and healthcare infrastructure can delay the widespread implementation of digital technologies.

The main challenge in robotics rehabilitation is the high cost and accessibility issues, particularly in low- and middle-income countries. Additionally, older adults have physical and physiological limitations, e.g., age-related fatigue may limit the duration and intensity of the training session. Older adults also exhibit cognitive decline and memory deficits, making it difficult for them to understand instructions and interact with robots. Visual and auditory deficits further complicate the use of such systems. Patients may also exhibit fear and apprehension about using technology-based interventions. Effective training sessions also require technical knowledge and expertise. The hip-knee exoskeletons also increase metabolic costs due to their heavy weight. Another challenge related to wearable robotic devices is the lack of sufficient evidence on the clinical use of different device types across diverse clinical populations (Cha et al., 2025). The intervention dosage, in terms of frequency, intensity, and timing, still needs to be refined. This results in reduced adherence, safety, and effectiveness of robot-based treatment interventions. Though these challenges exist, there is still a window of opportunity to make these technologies more practical and accessible to a wider audience (Cha et al., 2025).

The challenges related to VR, AR, and exergaming include side effects such as motion sickness and mild, transient symptoms like nausea, eye discomfort, and disorientation (Dore et al., 2023). Older adults may find immersive VR dynamic activities, such as walking, stressful, and a familiarisation session before the intervention is needed to ensure their comfort and feasibility. Also, age-related vision loss becomes a challenge. However, this can be mitigated through adjustable visual fields and lighting, targeted visual cues during the game, and optical aids such as prisms should be used in AR to compensate for visual field deficits (Dore et al., 2023). Older adults may also find using controllers challenging, especially if they have limited exposure to technology. To overcome this, the newly developed systems incorporate hand-tracking technology to enable VR headset use without controllers (Dore et al., 2023). Successful implementation of VR technologies requires standardising intervention protocols and training clinicians. The heterogeneity of the VR interventions, participants, settings, and outcome measures, and the small sample size make it challenging to use VR clinically.

The main challenge with WDs is their limited validity and accuracy across healthy and clinical older adults. Like other technological innovations, cost and affordability also limit the widespread use of these WDs. For older adults, data privacy and security concerns, comfort, usability, and acceptance also become critical challenges. Limited interoperability between systems and high energy consumption when using IoT-enabled sensors are additional challenges. Despite the many advantages of telerehabilitation, several challenges limit its use. Digital literacy barriers, safety concerns such as the risk of falls during balance training, and age-related comorbidities can impede the use of telerehabilitation in older adults (Gamble et al., 2024). A scoping review by Kebede et al. mapped barriers and facilitators using the COM-B (Capability, Opportunity, Motivation-Behaviour) and Theoretical Domains Frameworks. According to this, capability-related barriers, such as reduced vision, hearing, dexterity, or cognitive function, can limit the effective use of rehabilitation technologies. Opportunity-related factors, including access to reliable internet, technical support, and user-friendly design, affect the engagement level of the older adults. Motivation, including perceived usefulness, confidence, and trust in technology, plays a convincing role in continued participation (Kebede et al., 2022).

Dispositional factors, including older adults' beliefs and attitudes towards technology, are another challenge that needs to be addressed. There is a strong attachment to daily routines, a lack of interest, and fear of using technological interventions, such as apprehension about online fraud, sudden device malfunctions, or making mistakes while using healthcare technologies, which can further hinder engagement and adoption of digital technologies (Bertolazzi et al., 2024). A recent rapid review examining ethics and equity challenges in telerehabilitation for older adults highlights significant concerns that must be addressed to ensure accessible and inclusive telerehabilitation practices. Although telerehabilitation offers multiple benefits for older adults' care, the review identifies socioeconomic, geographic, and racial disparities that affect its adoption and effectiveness. Older adults from lower-income backgrounds, rural regions, and minority communities face limited access to digital infrastructure and reliable internet connectivity. The review also highlights ethical issues, including data privacy, cybersecurity, informed consent, and safety monitoring, that remain insufficiently addressed in many telerehabilitation studies (Veras et al., 2025).

Despite advancements in Intelligent Assistive Technologies for ageing at home, similar challenges do exist with these systems as well. Because of the sensitive nature of the data and the need for continuous monitoring, challenges persist in ensuring privacy, security, and ethical compliance. Addressing the challenges and barriers during the design and evaluation of smart home technologies is important to ensure that these emerging technologies support ageing in place while respecting core ethical values.

5.6 Conclusion and Future Directions

Future digital interventions should be designed to meet the individual needs, preferences, and functional status of older adults. There is a need to design lightweight equipment and robotic devices. They should be designed to enable easy use in home and low-resource environments, addressing the challenges of affordability, ease of access, and use by older adults. Emerging technologies should empower independent older adults to use wearables for self-monitoring and lifestyle management across diverse care settings.

There is a need to design affordable, easy-to-use devices to support active, and healthy ageing. Infrastructure support should be provided to rural and underserved communities to ensure equitable access. Features such as gamification, social connectivity, reminders, and caregiver support should be added to ensure engagement and motivation of older adults.

Future development must include ethical frameworks for data privacy, consent, and security. In conclusion, technologies should empower older adults while safeguarding their autonomy and confidentiality.

References

Aldardour, A., & Alnammaneh, S. (2025). The role of immersive virtual reality in geriatric rehabilitation. *Rehabilitación, 59*(3), Article 100930.

Alruwaili, M. M., Shaban, M., & Elsayed Ramadan, O. M. (2023). Digital health interventions for promoting healthy aging: A systematic review of adoption patterns, efficacy, and user experience. *Sustainability, 15*(23), 16503.

Banyai, A. D., & Brişan, C. (2024). *Robotics in Physical Rehabilitation: Systematic Review, 12*(17), 1720.

Bertolazzi, A., Quaglia, V., & Bongelli, R. (2024). Barriers and facilitators to health technology adoption by older adults with chronic diseases: An integrative systematic review. *BMC Public Health, 24*(1), 506. https://doi.org/10.1186/s12889-024-18036-5

Bueno, G. A. S., do Bomfim, A. D., Campos, L. F., Martins, A. C., Elmescany, R. B., Stival, M. M., Funghetto, S. S., & de Menezes, R. L. (2024). Non-invasive neuromodulation in reducing the risk of falls and fear of falling in community-dwelling older adults: Systematic review. *Frontiers in Aging Neuroscience, 15*, 1301790.

Ceyssens, C., Verhulst, N., & Willems, K. (2025). Enhancing the well-being of older adults through augmented reality: A systematic literature review and future research agenda. *International Journal of Consumer Studies, 49*(4), Article e70080.

Cha, J. M., Hong, J., Yoo, J., & Rha, D. (2025b). Wearable robots for rehabilitation and assistance of gait: A narrative review. *Annals of Rehabilitation Medicine, 49*(4), 187–195. https://doi.org/10.5535/arm.250093

Chang, H., Song, Y., & Cen, X. (2022). Effectiveness of augmented reality for lower limb rehabilitation: A systematic review. *Applied Bionics and Biomechanics, 2022*, 4047845. https://doi.org/10.1155/2022/4047845

Cidre, E. (2025). Active (and healthy) ageing in the built environment. *Journal of Urban Design, 30*(2), 141–142.

Correa, F. I., Carneiro Costa, G., Leite Souza, P., Marduy, A., Parente, J., Ferreira da Cruz, S., de Souza Cunha, M., Beber Freitas, M., Correa Alves, D., & Silva, S. M. (2023). Additive effect of transcranial direct current stimulation (tDCS) in combination with multicomponent training on elderly physical function capacity: A randomized, triple blind, controlled trial. *Physiotherapy Theory and Practice, 39*(11), 2352–2365.

Corregidor-Sánchez, A. I., Segura-Fragoso, A., Rodríguez-Hernández, M., Jiménez-Rojas, C., Polonio-López, B., & Criado-Álvarez, J. J. (2021). Effectiveness of virtual reality technology on functional mobility of older adults: Systematic review and meta-analysis. *Age and Ageing, 50*(2), 370–379.

Das, A., & Dhillon, P. (2023). Application of machine learning in measurement of ageing and geriatric diseases: A systematic review. *BMC Geriatrics, 23*(1), 841.

Dawson, R., Oliveira, J. S., Kwok, W. S., Bratland, M., Rajendran, I. M., Srinivasan, A., Chu, C. Y., Pinheiro, M. B., Hassett, L., & Sherrington, C. (2024). Exercise interventions delivered through telehealth to improve physical functioning for older adults with frailty, cognitive, or mobility disability: A systematic review and meta-analysis. *Telemedicine and E-Health, 30*(4), 940–950.

Dore, B., Gaudreault, A., Everard, G., Ayena, J. C., Abboud, A., Robitaille, N., & Batcho, C. S. (2023). Acceptability, feasibility, and effectiveness of immersive virtual technologies to promote exercise in older adults: A systematic review and meta-analysis. *Sensors, 23*(5), 2506.

Gamble, C. J., van Haastregt, J., van Dam van Isselt, E. F., Zwakhalen, S., & Schols, J. (2024). Effectiveness of guided telerehabilitation on functional performance in community-dwelling older adults: A systematic review. *Clinical Rehabilitation, 38*(4), 457–477. https://doi.org/10.1177/02692155231217411

Ghafurian, M., Wang, K., Dhode, I., Kapoor, M., Morita, P. P., & Dautenhahn, K. (2023). Smart home devices for supporting older adults: A systematic review. *IEEE Access, 11*, 47137–47158.

Guerra, B. M. V., Torti, E., Marenzi, E., Schmid, M., Ramat, S., Leporati, F., & Danese, G. (2023). Ambient assisted living for frail people through human activity recognition: State-of-the-art, challenges and future directions. *Frontiers in Neuroscience, 17*, 1256682.

Gustafsson, T., & Ulfhake, B. (2024). Aging skeletal muscles: What are the mechanisms of age-related loss of strength and muscle mass, and can we impede its development and progression? *International Journal of Molecular Sciences, 25*(20), 10932.

Høeg, E. R., Povlsen, T. M., Bruun-Pedersen, J. R., Lange, B., Nilsson, N. C., Haugaard, K. B., Faber, S. M., Hansen, S. W., Kimby, C. K., & Serafin, S. (2021). System immersion in virtual reality-based rehabilitation of motor function in older adults: A systematic review and meta-analysis. *Frontiers in Virtual Reality, 2*, Article 647993.

Hong, S., Baek, S.-H., Lai, M. K., Arumugam, T. V., & Jo, D.-G. (2024). Aging-associated sensory decline and Alzheimer's disease. *Molecular Neurodegeneration, 19*(1), 93.

Izquierdo, M., Ramírez-Vélez, R., & Singh, M. A. F. (2025). Integrating exercise and medication management in geriatric care: A holistic strategy to enhance health outcomes and reduce polypharmacy. *The Lancet Healthy Longevity, 6*(9).

Ju, F., Wang, Y., Xie, B., Mi, Y., Zhao, M., & Cao, J. (2023). *The Use of Sports Rehabilitation Robotics to Assist in the Recovery of Physical Abilities in Elderly Patients with Degenerative Diseases: A Literature Review, 11*(3), 326.

Kebede, A. S., Ozolins, L.-L., Holst, H., & Galvin, K. (2022b). Digital engagement of older adults: Scoping review. *Journal of Medical Internet Research, 24*(12), Article e40192. https://doi.org/10.2196/40192

Lee, K., Yi, J., & Lee, S.-H. (2024). Effects of community-based fall prevention interventions for older adults using information and communication technology: A systematic review and meta-analysis. *Health Informatics Journal, 30*(2), 14604582241259324.

Lee, S. H., & Yu, S. (2025). The impact of care robots on older adults: A systematic review. *Geriatric Nursing, 65*, Article 103507.

Leghissa, M., Carrera, Á., & Iglesias, C. A. (2023). Machine learning approaches for frailty detection, prediction and classification in elderly people: A systematic review. *International Journal of Medical Informatics, 178*, Article 105172.

Li, Z., Zhang, Z., Ren, Y., Wang, Y., Fang, J., Yue, H., Ma, S., & Guan, F. (2021). Aging and age-related diseases: From mechanisms to therapeutic strategies. *Biogerontology, 22*(2), 165–187.

Lim, J.-Y. (2025). Core principles and structures of geriatric rehabilitation: A narrative review. *Ewha Medical Journal, 48*(4), Article e57.

Lingampally, P. K., Ramanathan, K. C., Shanmugam, R., Cepova, L., & Salunkhe, S. (2024). Wearable assistive rehabilitation robotic devices—A comprehensive review. *Machines, 12*(6), 415.

Liu, M., Zhou, K., Chen, Y., Zhou, L., Bao, D., & Zhou, J. (2022). Is virtual reality training more effective than traditional physical training on balance and functional mobility in healthy older adults? A systematic review and meta-analysis. *Frontiers in Human Neuroscience, 16*, Article 843481. https://doi.org/10.3389/fnhum.2022.843481

Ma, B., Yang, J., Wong, F. K. Y., Wong, A. K. C., Ma, T., Meng, J., Zhao, Y., Wang, Y., & Lu, Q. (2023). Artificial intelligence in elderly healthcare: A scoping review. *Ageing Research Reviews, 83*, Article 101808.

Ma, L., & Chan, P. (2020). Understanding the physiological links between physical frailty and cognitive decline. *Aging and Disease, 11*(2), 405.

Maranesi, E., Amabili, G., Cucchieri, G., Bolognini, S., Margaritini, A., & Bevilacqua, R. (2022). Understanding the acceptance of ioT and social assistive robotics for the healthcare sector: A review of the current user-centred applications for the older users. *Internet of Things for Human-Centered Design: Application to Elderly Healthcare*, 331–351.

Marzola, P., Melzer, T., Pavesi, E., Gil-Mohapel, J., & Brocardo, P. S. (2023). Exploring the role of neuroplasticity in development, aging, and neurodegeneration. *Brain Sciences, 13*(12), 1610.

Mikhailova, V. (2025). A scoping literature review on older adults' experiences with wearable augmented reality: Opportunities and challenges for successful aging. *International Journal of Human–Computer Interaction*, 1–22.

Nairn, B., Tsakanikas, V., Gordon, B., Karapintzou, E., Kaski, D., Fotiadis, D. I., & Bamiou, D.-E. (2025). Smart wearable technologies for balance rehabilitation in older adults at risk of falls: Scoping review and comparative analysis. *JMIR Rehabilitation and Assistive Technologies, 12*, Article e69589. https://doi.org/10.2196/69589

Olawade, D. B., Adeleye, K. K., Egbon, E., Nwabuoku, U. S., David-Olawade, A. C., Boussios, S., & Vanderbloemen, L. (2025). Enhancing home rehabilitation through AI-driven virtual assistants: A narrative review. *Annals of Translational Medicine, 13*(5), 61.

Olmedo-Aguirre, J. O., Reyes-Campos, J., Alor-Hernandez, G., Machorro-Cano, I., Rodriguez-Mazahua, L., & Sanchez-Cervantes, J. L. (2022). Remote healthcare for elderly people using wearables: A review. *Biosensors, 12*(2), 73.

Peng, D., Zhang, S., Zhang, Y., Wu, H., Zou, X., Zhang, G., Ma, Y., Ma, T., & Wang, L. (2025). Effects of unstable resistance training combined with high-definition transcranial direct current stimulation on proprioception, balance and fall risks in the elderly: A randomized controlled trial study. *Experimental Gerontology*, 112843.

Preitschopf, A., Holstege, M., Ligthart, A., Groen, W., Burchell, G., Pol, M., & Buurman, B. (2023). Effectiveness of outpatient geriatric rehabilitation after inpatient geriatric rehabilitation or hospitalisation: A systematic review and meta-analysis. *Age and Ageing, 52*(1), afac300.

Robinson, F., & Nejat, G. (2022). An analysis of design recommendations for socially assistive robot helpers for effective human-robot interactions in senior care. *Journal of Rehabilitation and Assistive Technologies Engineering, 9*, 20556683221101388.

Rodriguez-Almagro, D., Achalandabaso-Ochoa, A., Ibanez-Vera, A. J., Gongora-Rodriguez, J., & Rodriguez-Huguet, M. (2024). *Effectiveness of Virtual Reality Therapy on Balance and Gait in the Elderly: A Systematic Review, 12*(2), 158.

Rodríguez-Almagro, D., Achalandabaso-Ochoa, A., Ibáñez-Vera, A. J., Góngora-Rodríguez, J., & Rodríguez-Huguet, M. (2024). Effectiveness of Virtual Reality Therapy on Balance and Gait in the Elderly: A Systematic Review. *Healthcare, 12*(2), 158. https://doi.org/10.3390/healthcar e12020158

Salma, I., Testa, D., Mpoy, J., Perez-Torrents, J., Rehault, J., Cabanes, E., & Minvielle, E. (2025). Remote monitoring system for older adults at risk for complications: A scoping review. *BMC Geriatrics, 25*(1), 766.

Santana Muñoz, D., Lorca Navarro, M., Araya Orellana, E., Moscoso Aguayo, P., & Martínez Huenchullán, S. (2024). Effect of training with immersive virtual reality on the risk of falling in the elderly: A systematic review. *Rehabilitacion (Madr)*, 100857–100857.

Shah, S. M., Aljawarneh, M. M., & Akbar, S. (2025). Assistive living in IoT-based smart home systems for aging in place: A review. *VFAST Transactions on Software Engineering, 13*(2), 83–98.

Sharafi, V., & Mohammadyari, Z. (2025). Applications of artificial intelligence in geriatric care management: a meta-synthesis approach. *Iranian Journal of Ageing, 20*(3), 400–421.

Stavropoulos, T. G., Papastergiou, A., Mpaltadoros, L., Nikolopoulos, S., & Kompatsiaris, I. (2020). IoT wearable sensors and devices in elderly care: A literature review. *Sensors, 20*(10), 2826.

Sumner, J., Lim, H. W., Chong, L. S., Bundele, A., Mukhopadhyay, A., & Kayambu, G. (2023). Artificial intelligence in physical rehabilitation: A systematic review. *Artificial Intelligence in Medicine, 146*, Article 102693.

Teixeira, E., Fonseca, H., Diniz-Sousa, F., Veras, L., Boppre, G., Oliveira, J., Pinto, D., Alves, A. J., Barbosa, A., & Mendes, R. (2021). Wearable devices for physical activity and healthcare monitoring in elderly people: A critical review. *Geriatrics, 6*(2), 38.

Tenchov, R., Sasso, J. M., Wang, X., & Zhou, Q. A. (2023). Aging hallmarks and progression and age-related diseases: A landscape view of research advancement. *ACS Chemical Neuroscience, 15*(1), 1–30.

Tian, Y. J., Felber, N. A., Pageau, F., Schwab, D. R., & Wangmo, T. (2024). Benefits and barriers associated with the use of smart home health technologies in the care of older persons: A systematic review. *BMC Geriatrics, 24*(1), 152.

Valenzuela, P. L., Saco-Ledo, G., Morales, J. S., Gallardo-Gómez, D., Morales-Palomo, F., López-Ortiz, S., Rivas-Baeza, B., Castillo-García, A., Jiménez-Pavón, D., & Santos-Lozano, A. (2023). Effects of physical exercise on physical function in older adults in residential care: A systematic review and network meta-analysis of randomised controlled trials. *The Lancet Healthy Longevity, 4*(6), e247–e256.

Veras, M., Auger, L.-P., Sigouin, J., Gheidari, N., Nelson, M. L., Miller, W. C., Hudon, A., & Kairy, D. (2025). Ethics and equity challenges in telerehabilitation for older adults: Rapid review. *JMIR Aging, 8*(1), Article e69660. https://doi.org/10.2196/69660

Wang, J., Li, Y., Yang, G.-Y., & Jin, K. (2024). Age-related dysfunction in balance: A comprehensive review of causes, consequences, and interventions. *Aging and Disease, 16*(2), 714.

Wang, J., Liang, Y., Cao, S., Cai, P., & Fan, Y. (2023). Application of artificial intelligence in geriatric care: Bibliometric analysis. *Journal of Medical Internet Research, 25*, Article e46014.

Wang, Z., Yang, Z., & Dong, T. (2017). A review of wearable technologies for elderly care that can accurately track indoor position, recognize physical activities and monitor vital signs in real time. *Sensors, 17*(2), 341.

Weizman, Y., Tirosh, O., Beh, J., Fuss, F. K., & Pedell, S. (2021). Gait assessment using wearable sensor-based devices in people living with dementia: A systematic review. *International Journal of Environmental Research and Public Health, 18*(23), 12735.

Wicks, M., Dennett, A. M., & Peiris, C. L. (2023). Physiotherapist-led, exercise-based telerehabilitation for older adults improves patient and health service outcomes: A systematic review and meta-analysis. *Age and Ageing, 52*(11), afad207.

Wong, E. K., Hoang, P. M., Kouri, A., Gill, S., Huang, Y. Q., Lee, J. C., Weiss, S. M., Daniel, R., McGowan, J., & Amog, K. (2024). Effectiveness of geriatric rehabilitation in inpatient and day hospital settings: A systematic review and meta-analysis. *BMC Medicine, 22*(1), 551.

World Health Organization. (2023). *Progress report on the United Nations decade of healthy ageing, 2021–2023*. World Health Organization.

World Health Organization. (2025). *Ageing and health*. Ageing and Health. https://www.who.int/news-room/fact-sheets/detail/ageing-and-health

Yamout, Y., Yeasar, T. S., Iqbal, S., & Zulkernine, M. (2023). Beyond smart homes: An in-depth analysis of smart aging care system security. *ACM Computing Surveys, 56*(2), 1–35.

Yang, X., Li, S., Xu, L., Liu, H., Li, Y., Song, X., Bao, J., Liao, S., Xi, Y., & Guo, G. (2024). Effects of multicomponent exercise on frailty status and physical function in frail older adults: A meta-analysis and systematic review. *Experimental Gerontology, 197*, Article 112604.

Yixiao, C., Hui, S., Quhong, S., Xiaoxi, Z., & Jirong, Y. (2025). A review of utility of wearable sensor technologies for older person frailty assessment. *Experimental Gerontology, 200*, Article 112668.

Zhang, H., Xu, J., Lu, X., & Xu, M. (2025). Wearable devices in elderly chronic disease management: A qualitative study of barriers and facilitators. *Journal of Nursing Management, 2025*(1), 1278057.

Chapter 6
Women's Health Disabilities and Technology-Enabled Rehabilitation

Huma Riaz ⓘ

Abstract Women experience a distinct spectrum of health conditions leading to disability, ranging from musculoskeletal, hormonal, reproductive, and oncological domains. The complex set of physiological, hormonal, and social determinants disproportionately contributes to long-term disability in women. Prevalent physical disability conditions, such as postmenopausal osteoporosis and fragility fractures, post-mastectomy lymphedema, and post-hysterectomy CPP, have a distinct set of impairments and functional limitations. The hormonal variations, reproductive surgeries, central sensitisation, and altered pain pathways are the underpinning pathophysiological mechanisms for disability. This chapter provides readers with an overview of the role of emerging technology-based rehabilitation solutions in addressing disability-related women's health conditions. VR and digital health platforms, wearable sensor-based monitoring, and whole-body vibration devices are used for balance control, fall prevention, bone remodelling stimulation, and fracture rehabilitation support. AI-based personalised rehab plans, VR, robotics with upper-limb exoskeletons, smart textiles, and digital platforms are effective in treating breast cancer-related lymphedema. Multimodal TR systems, electromyography BF, and digital apps support pelvic floor rehab in diverse dysfunctions. This chapter concludes with the limitations in technology applications and future directions in disability management.

Keywords Hysterectomy · Lymphedema · Menopause · Osteoporosis · Pelvic pain · Women's health

H. Riaz (✉)
Faculty of Rehabilitation and Allied Health Sciences, Riphah International University, Islamabad, Pakistan
e-mail: huma.riaz@riphah.edu.pk

6.1 Introduction

Women experience a distinct spectrum of health conditions leading to disability, ranging from musculoskeletal, reproductive, autoimmune, and oncological domains. The complex set of biological, physiological, hormonal, and sociocultural determinants disproportionately contributes to long-term disability in women. The longer duration of disability in women is also linked to increased longevity. The intersectionality of gender, disability, and poverty seems to worsen the health and social outcomes in women, as they face barriers in rehabilitation, societal participation, and employment. Physical disabilities (PD) in women constitute a significant public health problem that contributes to a greater burden of disability in them as compared to men. In the context of designing effective and equitable technology-enabled rehabilitation strategies and solutions, it is essential to review epidemiological trends of physical disability and prevalent health conditions among women.

6.2 Global Burden and Prevalence of Disability Conditions in Women

Disability is a gendered experience, with the global prevalence higher in women versus men (approx. 18% vs 14%) in almost all countries, across all age groups. But this gender gap widens with age. Globally, women constitute >52% of persons with disabilities, with the higher burden of disability leading to chronic health conditions and increased life expectancy being the major disability drivers (World Health Organization, 2022). Disability prevalence is higher among older women >60 years, disproportionately affected by musculoskeletal and pain disorders; they live longer but have more years lived with disabilities (YLDs) (World Health Organization, 2023). Ageing and chronic disease convergence make older women the largest subgroup of disabled adults globally.

As women transition across the lifespan, the burden of physical disabilities is increasingly shaped by age and hormone depletion-related bone loss, called post-menopausal osteoporosis, and the related risk of fragility fractures. In the continuum, the long-term sequelae of various medical and surgical interventions, like breast cancer-related or post-mastectomy lymphedema and post-gynaecological surgery or post-hysterectomy CPP, are encountered in women's life course as disabling conditions.

6.2.1 *Postmenopausal Osteoporosis and Fragility Fractures*

Disability prevalence in women also varies with lifespan, reflecting both the physiological changes and cumulative exposure to health risks. Postmenopausal osteoporosis (PMOP) is a skeletal disorder affecting women increasingly after their midlife. It is characterised by microarchitectural deterioration in bone tissue and reduced bone mineral density (BMD), caused by an imbalance in osteoblastic and osteoclastic activity, resulting in decreased bone strength, increased bone fragility, and an associated increased risk of fractures. Ageing leads to a gradual imbalance between bone resorption and new bone formation, which is accelerated by a rapid decline in female reproductive hormones in the first few years after menopause. Osteoporosis represents one of the prevalent causes of PD, as the structural skeletal compromise leads to significant spinal postural deformities, chronic pain, fear of falls, and finally loss of functional independence (Lorentzon et al., 2022).

Approximately 200 million women worldwide are affected by osteoporosis, defined as a femoral neck BMD T-score of -2.5 SD or below. As BMD declines with age, its proportion declines as well, affecting 10%, 20%, 40%, and 2/3 of women at ages 60, 70, 80, and 90 years, respectively (Lorentzon et al., 2022). Most fractures experienced by postmenopausal women (PMW) are low-trauma fractures or fragility fractures and fall in the osteoporotic category. Globally, one in three women aged 50 or older will sustain an osteoporotic fracture in their remaining lifetime. In general, women experience fractures twice as often as men, and in the case of hip fractures, the rate is nearly 75% (Jordan & Cooper, 2002). One year after hip fracture, 40% of women are unable to walk independently, 60% have difficulty with activities of daily living (ADLS), and 80% have difficulty shopping and driving. Common sites of fragility fractures include the hip, spine, and distal radius. In 2001, approximately 9 million new osteoporotic fractures were reported worldwide, 51% in Western countries and the remainder in Southeast Asia. The common clinical sites reported were 1.7 million in the forearm, 1.4 million in the vertebral region, and 1.6 million in the hip region (Johnell & Kanis, 2006).

Fragility fractures of clinical sites, such as the hip and spine, constitute a substantial public health problem, and a significant decline in body function leads to long-term consequences, including chronic pain, disability, decreased quality of life, morbidity, and mortality. In women, the incidence of vertebral and hip fractures rises sharply after 60 and 70 years, respectively (Sambrook & Cooper, 2006). Women with osteoporosis become progressively disabled, and those having mobility limitations and pain become more osteoporotic, and the disability cycle perpetuates. The 2021 global epidemiology study, covering data from 1990 to 2021 and 204 countries, reports that low BMD was responsible for 7.76 million disability-adjusted life years (DALYs) and 219,552 deaths among PMW worldwide (Liang et al., 2025). In high-income countries, the direct annual economic cost of PMOP and related fracture treatment is estimated at 5000–6500 billion USD. And these estimates do not include loss of productivity or long-term caregiving costs, so the total societal cost is even larger (Kemmak et al., 2020).

6.2.2 Breast Cancer-Related or Post-mastectomy Lymphedema

Breast cancer-related, or post-mastectomy lymphedema, is a chronic, progressive, often irreversible, and potentially debilitating consequence in breast cancer survivor women. It is characterised by impaired accumulation of protein-rich lymphatic fluid in the interstitial tissues of the upper limb (sometimes extending to the adjoining breast and trunk), resulting from mechanical limitations of the lymphatic system following breast cancer treatment. This imbalance arises from either overloading or diminishing the lymphatic system's drainage capacity, leading to oedema and tissue structural alterations (Bakar, 2025). It is not a transient, postsurgical complication but a chronic, often progressive disability requiring lifelong management. With long-term treatment, the direct cost of treatment lymphedema and associated disability and indirect cost of work loss, productivity, and social disability, brings significant financial toxicity (Beck & Lizarraga, 2024).

Breast Cancer-Related Lymphedema (BCRL) affects approximately 20% of women. With the advances in cancer treatments, the life expectancy and prevalence of lymphedema will continue to rise and affect the quality of life of breast cancer survivors. High-quality evidence suggests that one in five breast cancer survivor women experience post-mastectomy lymphedema, with increasing risk over the years after treatment. A 2013 systematic review and meta-analysis of studies from higher-income countries reports an incidence of unilateral lymphedema ranging from 8.4% to 21.4%. The pooled incidence of lymphedema is 16.6%, compared to a pooled estimate of 21.4% when restricted to prospective cohort studies. After axillary surgery, the longitudinal cohort data show a rising trend in the cumulative incidence of arm lymphedema over time: 13.5%, 30.2%, and 41.1% at 2, 5, and 10 years, respectively. The associated factors were listed as radiotherapy, chemotherapy, seroma, obesity, and cancer stage (DiSipio et al., 2013; Li et al., 2017). In lower-middle-income countries, the incidence and prevalence of post-mastectomy lymphedema are generally higher, widely ranging from 0.4% in Papua New Guinea to 92.5% reported in a Brazilian study. The overall pooled incidence was 21%, with the lowest reported incidence, 5.9%, in Romania, and the highest, 56.7%, in an Indian study (Torgbenu et al., 2020). There is wide variability and heterogeneity in reporting prevalence and incidence estimates of arm lymphedema after taking breast cancer treatment, due to variations in measurement methods, stage of breast cancer, and the treatment received.

6.2.3 Post-hysterectomy Chronic Pelvic Pain

Chronic pelvic pain (CPP) is a debilitating, multifactorial clinical condition, primarily affecting women (1 in 4) and estimated to be 25% worldwide. American College of Obstetrics and Gynaecology defines it as pain in anatomical regions (pelvic organs/

structures) and of duration >6 months; by International Association for Study of Pain, emphasising specific pain syndromes of shorter duration (3 months); by World Health Organization (WHO) ICD-11, classifying into chronic primary and secondary pain; and by RCOG, characterising CPP as a symptom. It poses a significant health and economic burden on women's lives, affecting their physical, mental, and social well-being (Villegas-Echeverri et al., 2025). Hysterectomy is the most common gynaecological surgical procedure performed on women all over the world, with a low risk of major complications (Kovac, 2014). It can be performed laparoscopically, vaginally, or by laparotomy. Traditionally, the abdominal hysterectomy procedure was the most common technique, but now it is limited to a large uterus. In modern times, less invasive procedures such as vaginally or laparoscopic hysterectomy are practised. In the USA, an estimated 12% hysterectomies are done annually, with the primary indication of endometriosis, and most women recover from the experience of CPP and return to life. Although hysterectomy is touted as a definitive intervention for CPP, previous studies inform that some women, 1 out of 4, experience discomfort, pain, and morbidity after the surgery. Post-hysterectomy CPP persists for more than 3 months after surgery, with the reported incidence of 27%, and it may peak to 40% in women with pre-existing CPP conditions (Misal et al., 2025). Predictors and risk factors have been suggested in literature as young age, significant perioperative pain, genetic susceptibility, pre-existing centralised pain, and psychological factors including anxiety, depression, and pain catastrophising (Kehlet et al., 2006). Likewise, a SR found mixed or uncertain associations with endometriosis, surgical approach, and the effects of ovarian conservation during the procedure (Misal et al., 2025).

6.2.4 Socioeconomic and Cultural Determinants of Disability in Women

As per WHO's biopsychosocial framework of functioning and disability, the PD among women is not solely a matter of biological vulnerability, reproductive events, and clinical exposure, but is a socially mediated phenomenon. The socioeconomic status, cultural norms, and health system inequalities profoundly shape their experience of living with disability. The gender barriers, such as lesser access to rehabilitation services, amplify the disability impact. Furthermore, economic dependence and poverty worsen restrictions on societal participation (World Health Organization, 2022). As per UNICEF, the WHO Global report on children with developmental disabilities, the incidence of female disabilities starts rising from adolescence as compared to their male counterparts. With their early caregiving roles, they have a high vulnerability to educational exclusion, which has long-term social and functional impacts (UNICEF, 2023). The UNSD disability statistics report informs us that women with disabilities have lower educational attainment and employment rates as compared to men, which places them at lifelong socioeconomic disadvantage and greater participation restrictions (UN Affairs, 2025). So, the incidence and lifelong burden of physical disabilities, such as osteoporosis, lymphedema, and CPP, are more prevalent or selective in women and are largely linked to global disparities, including but not limited to early diagnosis, access to preventive care, surgical safety, and rehabilitation services.

PMOP is largely a preventable skeletal disorder. Still, the sociocultural and economic inequalities faced by women intensify their burden along with those of fragility fractures, transforming it into a major source of PD. The intersection of socioeconomic, cultural, and systemic factors explains why women are disproportionately affected. Evidence consistently explains that there is reduced access to preventive BMD screening, particularly through dual energy X-ray absorptiometry, in racial, ethnic minorities, and socioeconomically disadvantaged women, leading to delayed disease identification and untreated loss of bone density (Calikyan et al., 2023; Ruiz-Esteves et al., 2022). The disability trajectory is further shaped by inequalities in health system capacity, as delays in the initiation of fracture surgeries and post-fracture services further contribute to disability in older women (Chandran et al., 2026). Health-seeking behaviour is affected by gender roles and cultural norms, taking osteoporosis as a normal and inevitable process of ageing, limiting the ageing women's access to screening, prevention, and rehabilitation services (Onizuka & Onizuka, 2024). Quantitative and qualitative literature reports that women with socioeconomic constraints are less engaged in bone-promoting health behaviours; inadequate knowledge and health literacy, provider communication gap, and misconceptions about fracture risk are the contributors (Dabas & Singh, 2024; des Bordes et al., 2020). Reviews of findings emphasise the need to strengthen health systems and adopt equity-oriented bone prevention strategies to mitigate disability from fragility fractures (Ali et al., 2024; Biz et al., 2024; Tagaev et al., 2025).

Women with BCRL of diverse racial and ethnic backgrounds face specific barriers and disparities in diagnosis and intervention accessibility. A recent SR reports that non-white women of younger age, rural background, with low education and income are at risk of self-reporting BCRL rather than being diagnosed by a physician. Those women with low socioeconomic status are at greater risk of breast cancer survivorship support and insufficient self-care education (Mattia et al., 2025). Based on individual needs or risks, breast cancer survivors should advocate for BCRL education. Social support and behaviour, including the essential role of family members and meeting with other affected women, are key factors in strengthening the self-management behaviour of lymphedema in breast cancer survivors. Contrary to this, inadequate support from the healthcare system, such as deficient medical resources, lack of follow-up, and an insufficient health insurance system, can make it difficult to maintain self-care behaviours (Perdomo et al., 2023; Wang et al., 2025). Breast cancer risk, survival, and development of lymphedema are also determined by consistent socioeconomic inequalities, including educational level, race, income level, and insurance type (Stout et al., 2024).

Women's pelvic health or pain is a globally essential subject, and its consequences are personal, intimate, and varied. The findings of a mixed-method SR demonstrate that pelvic pain symptoms are mostly perceived by women as stigmatised or non-urgent, identify major barriers to help-seeking as caregiving and job roles, time and cost constraints, and limited healthcare accessibility. These barriers, along with gendered expectations and cultural norms, make them realise that their symptoms are normal and trivial. Most women delay reporting symptoms out of fear of being unrecognised by clinicians (Jouanny et al., 2024). Women with lower socioeconomic status are less likely to receive adequate pain management and are at increased risk of persistent postsurgical pain syndromes (Thurston et al., 2023). A scoping review has identified prescription-based disparities, showing that despite higher pain reporting, female sex, lower socioeconomic status, and racial minority identity are linked with lower opioid administration, which directly affects their recovery and functional outcomes (Snell et al., 2025). Women with low educational level, poor health literacy, and insufficient disability awareness take postoperative pain as an inevitable consequence of surgery, so limited communication with the healthcare providers and inadequate understanding of postoperative pain management contribute to delayed identification and treatment of pain (Jouanny et al., 2024).

Collectively, these assumptions underscore that post-hysterectomy pain is not merely a biomedical outcome of surgical procedure but a socially mediated PD which is influenced by inequities in sociocultural norms and values, healthcare access, and responsiveness.

6.3 Pathophysiological Mechanisms of Disability Conditions

The pathophysiological foundations of disability prevalent women's health conditions explain the underlying biological, hormonal, and physiological mechanisms that translate the diseases or disorders into long-term physical impairments. The complex interplay of hormonal transitions, sex-specific anatomical, and surgical factors and altered pain processing pathways explains how structural and functional impairments lead to activity limitation and societal restrictions among women.

6.3.1 Hormonal Influences on the Musculoskeletal System and Function

The musculoskeletal health of women is largely regulated by oestrogen; its rapid decline during menopause is associated with changes in muscle, bone, connective tissue, ligaments, tendons, and joints (Takahashi & Johnson, 2015). The musculoskeletal syndrome of menopause is characterised by bone loss, sarcopenia, tendinopathies, joint pain, and diminished physical function. New-onset or worsening musculoskeletal pain cannot be explained solely by ageing (Wright et al., 2024).

The oestrogen decline at menopause accelerates bone loss, due to an imbalance of osteoclastic and osteoblastic activity, leading to osteoporosis and related fragility fractures (Ji & Yu, 2015). Likewise, muscle tissue expresses oestrogen receptors; its sudden decline compromises muscle strength, mass, and physical performance. Additionally, reduced oestrogen levels impair the muscle repair mechanism and enhance fatigue. Like muscles, tendons and ligaments are also oestrogen-sensitive; their decline is linked to altered mechanical properties, reduced cross-sectional area, and disorganisation of collagen fibres, increasing injury risk and decreasing functional capacity in women. Oestrogen deficiency is also associated with elevated pro-inflammatory cytokines, which may sensitise nociceptive pathways. The hormone also modulates pain; its decline contributes to altered pain pathways and central sensitisation in women (Ji & Yu, 2015).

Skeletal fragility due to osteoporosis is a structural impairment, a common cause of fractures at the hip, spine, and distal arm. Even with or without the fracture, women transit from independence to disability due to pain, postural deformity at the spine, immobilisation, and fear of movement. Due to early fatigue and reduced physical capacity, there is an increased risk of falls, resulting in impaired basic physical functions such as sit-to-stand, stair climbing, walking, and household activities. Structural alterations in tendons, ligaments, and joints impair joint stability and force transmission, making high-demand tasks such as lifting, carrying, and prolonged standing a great challenge for women. The altered central pain processing due to oestrogen decline is another hallmark of chronic disability. The amplified pain sensitisation impairs motor control, enhances fear avoidance behaviours, and reduces tolerance to movement (Ji & Yu, 2015).

6.3.2 Reproductive Surgeries as Drivers of Disability

In women, breast cancer and reproductive surgeries, such as hysterectomy, directly present structural and functional impairments, alterations in physical functions that follow participation restrictions.

Breast cancer survivors are at great risk of developing BCRL. Evidence has identified stronger associations with higher BMI, larger dissected lymph nodes, the extent of surgical procedure (e.g., total mastectomy), larger radiotherapy area, certain chemotherapy agents (e.g., taxane-based drugs), and a sedentary lifestyle, all of which are associated with BCRL (Byun et al., 2021; DiSipio et al., 2013). Progressive lymphedema can cause long-term physical impairments and dysfunctions, such as reduced upper-extremity mobility and strength, handgrip strength, and gross and fine-motor skills, as well as sensory impairments, including pain, heaviness, and tingling, leading to difficulty performing ADLS, self-care, and household tasks. It also enhances mental health and psychosocial burden by body image disturbance, social stigma and withdrawal, anger, anxiety, depression, guilt, and loss of perceived body control. These impairments contribute to delayed return to work, diminished work capacity, restrictions on social participation, role fulfilment, and the disability's economic consequences (Byun et al., 2021).

Post-hysterectomy CPP represents a clear aetiology for postsurgical pain that persists beyond the timeframe for tissue healing (>3 months). This woman's health-specific PD emerges as a constellation of various structural and functional impairments that affect their quality of life and ADLS. Almost 50–90% of CPP patients experience musculoskeletal pain and dysfunction. Structural changes in pelvic floor muscles, such as postsurgical scarring, altered muscle tissue compliance, myofascial trigger points, muscle shortening, fibrosis, or asymmetry, affect muscle strength (Lamvu et al., 2021). Adhesions in the pelvic and abdominal fascia, peripheral nerve irritation or entrapment (pudendal, ilioinguinal, genitofemoral, or iliohypogastric), and altered support structures (changes in ligamentous alignments and tensions) for pelvic organs, after removal of the uterus, affect bowel, bladder, and vaginal support (Moka et al., 2025). Similarly, altered physiological and neuromuscular functions bring functional impairments. These could be pelvic floor muscle dysfunction in the form of hypo- or hypertonicity, poor recruitment, and coordination during functional tasks such as coughing, sneezing, or postural changes (Pastore & Katzman, 2012). Gradually, reduced strength in the trunk, hip, and pelvic stabilisers limits daily activities, leads to guarded movement patterns, and results in fear-avoidant behaviours. There could be persistent pain processing dysfunction like central sensitisation with amplified pain responses and ongoing nociceptive, nociplastic, or neuropathic pain. Alteration in the pelvic organ support system could present as bowel or bladder dysfunction or sexual pain disorders, accompanied by altered sensations in the pelvic region (Cheong et al., 2014).

6.3.3 Chronic Pain, Altered Pain Pathways, and Central Sensitisation

There is gender disparity in the prevalence and presentation of chronic pain; globally, 20% population experiences it, and over 50% are women. Pain perception differs by sex, as women experience more chronic pain and greater disability. Female sex is a risk factor for chronic pain and its severity. Sociocultural factors, such as gender roles and healthcare access, further exacerbate this disparity. Women have a higher risk of developing chronic pain after an acute episode, such as surgery (Dance, 2019; Greenspan et al., 2007).

Hormonal changes throughout women's life span are reported to be responsible for variability in pain perception, sensitivity, and experience. Female hormones, oestrogen and progesterone, regulate pain through life cycles, with an increased rate of pain in girls at puberty as compared to boys. In contrast, these sex differences in chronic pain disappear as people age. The fluctuations in oestrogen during monthly cycles and the abrupt decrease in menopausal transitions promote inflammation and amplify pain perception. Additionally, animal studies indicate that females use remarkably different neural pathways involving T cells to mediate pain hypersensitivity (Sorge et al., 2015). Reproductive transitions from the menstrual cycle, pregnancy, and menopause further affect pain perception. PMW experience more chronic pain due to osteoporosis. Psychosocial and emotional factors complicate pain, its management, and predict transition to the chronic phase. Anxiety, depression, and distress can alter pain modulation pathways in the central nervous system, heighten pain perception, and reduce pain tolerance (Stieger et al., 2025). Women have a higher level of pain catastrophising, magnification, and associated helplessness as compared to men.

Chronic pain is a universal problem that leads to changes in the structure and function of the central nervous system; it is associated with alterations in certain brain areas, including decreased connectivity and grey matter. Postsurgical (hysterectomy) chronic pain has two mechanistic pathways: inflammatory and neuropathic. Inflammatory pain originates in response to tissue injury after surgery; the chemical mediators released increase peripheral nociceptor sensitivity, contributing to peripheral sensitisation. Neuropathic pain is induced by damage to the nerves traversing the surgical pathway, which releases pro-inflammatory cytokines and neurotrophic factors, producing a paradoxical combination of pain and sensory loss.

Evidence states that in some cases, central sensitisation or nociplastic pain is the underlying mechanism for persistent postoperative pelvic pain after hysterectomy. It is likely referred to as the disordered processing of the nervous system, in which painful experiences occur in the absence of ongoing tissue damage (Misal et al., 2025). Additionally, the myofascial pain or pain arising from the pelvic floor and abdominal wall fascia and musculature appears to play a vital role in the pain experiences of women with CPP, and this may perpetuate ongoing dysfunctions and symptoms after hysterectomy (Yosef et al., 2016).

6.4　Emerging Technologies in the Rehabilitation of Postmenopausal Osteoporosis and Fragility Fractures

PMW experience several structural and functional impairments due to oestrogen-linked bone loss and related skeletal fragility fractures. These are bone pain, decreased muscle strength and performance, postural deformities, balance deficits, fear of falls, and impaired functional capacity. These impairments lead to a gradual decline in functional mobility and participation. The traditional approaches of disability management of osteoporosis, while effective, are augmented by growing rehabilitation technologies. The current body of evidence indicates that these innovative intervention methods aim to be more engaging and promote long-term adherence among disabled women, while also demonstrating high-quality monitoring and precision.

6.4.1 Virtual Reality-Based Rehabilitation Technologies

In the management of osteoporosis, Virtual Reality (VR) has a multipurpose role by optimising neuromuscular coordination, proprioceptive integration, and patient compliance with therapy. To perform coordinated and good quality exercise, patients use motor, visual, and haptic signals in a virtual environment, leading to enhanced neuromuscular coordination. Proprioceptive integration is improved by real-time visual feedback, which reduces the risk of falls and fractures. Game-based VR training is more fun and motivating than traditional training and increases patient adherence (Maden et al., 2025; Raffegeau et al., 2023).

Non-Immersive and Camera-Based VR Systems focus on real-time feedback, gamification, and patient adherence to treatment protocol. Such exercise platforms, such as Kinect-based or camera-driven, can enhance functional balance, physical performance, and quality of life. A recent SRMA evaluating the effects of Non-immersive VR training from seven randomised controlled trails (RCTs) involving $n = 299$ women with bone loss has concluded that it significantly improves BMD, postural balance, functional mobility, and QoL (Cortés-Pérez et al., 2025). The consequence of improved impairments has been decreased pain and fear of falling. An RCT was conducted using gamified, Kinect-based VR training on PMW with bone loss. The experimental group has received a 45-min structured VR training session 3 times a week for 24 weeks, combined with a 30-min walk. The control group was instructed to walk outdoors for only 30 min. After 12 and 24 weeks of intervention, the Kinect-based VR training group has shown significant improvements in multiple physical performance measures and QoL, demonstrating it as a novel, feasible, and alternative rehabilitation approach (Riaz et al., 2024b). Another RCT has demonstrated improvements in lumbar spine and femoral neck BMD, physical performance, QoL, and reduced fracture risk in PMW with osteopenia after 24 weeks of gamified exercises using Kinect technology. The intervention has enabled progressive resistance training for the women while providing real-time feedback, and the game-based interfaces have engaged them well and supported long-term adherence to the rehabilitation protocol (Riaz et al., 2024a).

Immersive VR and Exergaming Platforms using immersive environments for gamified exercise rehabilitation protocol with head-mounted displays are playing an effective role in postural and balance control, and fall prevention for PMW. Although there is a paucity of literature on VR applications for the rehabilitation of women with osteoporosis, a recent SR has analysed various technologies that improve physical metrics, such as gait and balance, in osteoporosis and other musculoskeletal conditions. The review has classified technologies into four groups: first, head-mounted displays (HMDs), which provide an audio-visual experience with fully immersive and augmented reality, to improve gait and balance in older adults (Thuilier et al., 2024). For example, Microsoft HoloLens for fall prevention using mixed-reality balance training, I-Visor FX601; video glasses using augmented reality-based Otago exercises have shown significant improvements in balance, gait, and fall efficacy; Accupix MyBud; and a VR-based treadmill training to improve strength and symmetry for fall prevention in older women (Jung et al., 2012). Second, the balance boards, like Nintendo Wii Fit, for improving dynamic balance and weight shifting, ultimately fall prevention therapy with added motivation (Wall et al., 2015). Third, motion-detecting cameras and screens that can capture body poses, natural movements, and adaptability without wearable sensors. To mention some, are Microsoft Kinect, an RGB-D sensor effective for posture recognition and balance training, VR Rehabilitation System (VRRS), which improves range of motion and balance, GestureTek IREX, an Immersive Rehabilitation Exercise system, Sony PlayStation 2 Eye Toy, a motion-tracking sensor that allows interaction with virtual objects to support body movement rehabilitation (Thuilier et al., 2024). Lastly, the review highlighted the usability of other devices and technologies, such as Youkicker, a VR lower-limb system to improve lower-limb muscle strength and balance, and Cyber-Touch Dataglove to improve manual dexterity and hand functioning. The review findings indicate that digital interventions are potentially engaging for patients with long-term needs and disabilities, as they help prevent boredom and physical strain and enhance motivation and adherence. Some have usability challenges, but the choice of technology depends on the patient's clinical needs and desired level of sensory immersion.

Another preliminary study investigating the benefits of VRT versus a conventional multimodal approach in PMW with osteoporosis concluded that VRT was better at facilitating motor learning and improving functional outcomes, as it significantly reduced centre of pressure, sway velocity, and errors during functional tasks after 18 sessions over 6 weeks. To confirm the promising results, large trials are recommended (Rezaei et al., 2023).

6.4.2 Digital Health Platforms, Mobile Applications, and Telerehabilitation

Post-fragility fracture care can be facilitated with Digital health technologies through various mechanisms. SR findings demonstrate that digital health-supported patient communication systems improve secondary fracture prevention by enabling more efficient diagnosis and adherence to interventions for patients with osteoporosis. These interventions include mobile devices, telehealth platforms, and patient portals to maintain continuity of care and support self-management (Gupta et al., 2022).

Home-based Telerehabilitation programs provide a feasible and effective environment to improve mobility and functional performance in osteoporosis and post-fracture rehabilitation, especially in settings where access to in-person rehab options is limited. Such digital platforms address participation-level restrictions by extending disability rehabilitation beyond clinical settings.

Table 6.1 Digital health interventions for osteoporosis and post-fragility fracture care: purpose, features, and clinical relevance

mHealth apps	Purpose, features, and clinical relevance
FRAX® App	Calculates 10-year probability of a major osteoporotic fracture, Early identification, Patient education, Timely referral for BMD testing
My Osteoporosis Journey	Patient education, self-management, rehab, and lifestyle modification adherence display test results and connect patients & specialist
Hip Fracture Risk Calculator	Risk assessment for postsurgical complications or mortality after hip fracture, integration of risk prediction into tailored rehab goals, plans, and discharge discussions
AACE Osteoporosis Treatment Algorithm App	Complete evidence-based pathway and guide for diagnosis and treatment of PMOP (as per AACE guidelines), the therapist can align rehab interventions with medical management
My Osteoporosis Manager	Patient-centred app: medication adherence, test results, side effects, and symptoms; generates charts and summaries for clinician review; therapist can track data for progress and adjust rehab plans

Mobile Health (mHealth) Applications are designed specifically for osteoporotic patients, providing education, physical activity tracking, nutritional guidance, medication reminders, and adherence support. Such applications, as shown in Table 6.1, empower patients and help prevent long-term disability (Gupta et al., 2022).

6.4.3 Artificial Intelligence and Digital Risk Assessment Tools

Artificial intelligence (AI)-based tools are rapidly being integrated into osteoporosis care, primarily for early detection and risk stratification. AI-based automated fracture imaging detection and analysis use deep learning models to identify hip and vertebral fractures and perform radiological analysis, achieving high diagnostic accuracy. Similarly, algorithm-based fracture prediction tools, such as FRAX®, QFracture, and Garvan, alongside BMD, are frequently used to identify early osteoporosis risk and plan rehabilitation pathways. Although not the gold-standard fracture assessment tools, a SR suggests these have the best discriminative ability (Beaudoin et al., 2019).

Despite advancements in AI-based diagnosis and risk assessment, machine learning models integrating clinical, imaging, and biomechanical data for personalised rehabilitation planning still need to be developed. A Danish study informs the development and validation of FREM, a prediction model for high-risk osteoporotic or hip fractures (Beaudoin et al., 2019). Its integration into the electronic healthcare system will automate fracture risk detection, reducing reliance on human effort.

6.4.4 Wearable Sensors and Devices

To monitor disability recovery and progress, the objective quantification of physical activity, movements, and balance through wearable sensors and motion analysis technologies is an innovative and feasible solution. Continuous assessment of gait cycle, fall risk, and adherence to the intervention protocol can be achieved using wearable physical activity monitors and inertial sensors. A feasibility clinical trial was conducted with older adults following hip fracture; physical activity and step counts were measured remotely using smartwatches. Digital literacy and patient engagement were the main limitations, underscoring the need for more RCTs with elderly friendly designs (Hewage et al., 2023).

Mechano-stimulation wearable devices expose the skeleton to low-level vibrations, which can improve muscle strength, balance, and physical performance, although the direct effects on BMD are inconsistent. A 12-month RCT showed improvement in lumbar spine BMD in PMW with low bone mass using a wearable belt that delivers low-intensity vibrations, compared with placebo in the control group, suggesting potential to prevent osteoporosis. Large trials are recommended to evaluate cost-effectiveness and long-term fracture outcomes (Bilek et al., 2023).

Sensor-based technologies play a vital role in osteoporosis rehabilitation by increasing assessment precision and enabling rehabilitation protocol adjustments based on continuous data. A cross-sectional study was conducted to understand gait patterns in PMW with distal radius fragility fractures, using footwear sensors (In-Shoe Inertial Sensor) to capture real-life gait data of women and identify fall risks, linked gait irregularities, and other mobility impairments. The identification of gait patterns supported personalised rehab and prevention strategies. Large-scale and interventional trials are needed to guide therapy (Yamamoto et al., 2024). Another study has used ground reaction force sensors with AI to assess fracture-healing progress and complication risks, using detailed biomechanical gait data. The results demonstrated the potential to predict fracture healing and associated complications, thereby facilitating personalised rehabilitation plans. The study recommended clinical validation of osteoporotic-specific fractures and their integration into clinical decision-making tools (North et al., 2024). Other wearable devices, such as WISE (wearable inertial sensors for exergames) and ankle- or wrist-worn accelerometers, have been found beneficial for monitoring physical activity and gait kinetics in the rehabilitation of PMW with low bone mass (Bethi et al., 2020; Li et al., 2022; Madansingh et al., 2019).

The effects of mechanotherapy and Sensor-Assisted Exercise Technologies were evaluated in another study involving patients following hip fracture surgery. The intervention provided was sensor-based treadmill training with biofeedback (BF), as well as well-structured and controlled physical exercise, to enhance post-fracture recovery. The sensors provided feedback for balance and gait correction. Positive results were reported, with enhanced biomechanical parameters (Marchenkova et al., 2024).

6.4.5 Whole-Body Vibration Therapy

Whole-Body Vibration (WBV) therapy, a novel and adjuvant non-invasive mechanical stimulation, represents another diversified approach to osteoporosis rehabilitation. Low-magnitude, high-frequency mechanical stimuli delivered via vibration platforms activate muscle spindles, thereby inducing tonic vibration reflexes. Vibration therapy provides anabolic mechanical signals to bone cells and the musculoskeletal system. A review of the literature has concluded that WBV therapy has positive effects on bone mass and postural control in PMW (Singh & Varma, n.d.). Similarly, meta-analyses' findings of 13 RCTs with $n = 783$ PMOP patients have demonstrated that WBV significantly increases lumbar spine and femoral neck BMD (Li et al., 2024).

6.4.6 Posturography Systems and Technologies

PMW commonly experience back pain and postural deformities. There is a gradual shortening of women's height and a kyphotic posture due to a loss of BMD in the spine, with compression of the anterior vertebral body. To evaluate postural control, stabilometric platforms are used to assess body sway, and posturometric examinations are performed to assess the centre of pressure. Such systems can assess the sensory contributions to balance via protocols involving visual deprivation, providing insights into modifiable factors responsible for balance deficits and fall risk (Cultrera et al., 2010).

6.4.7 Spinal Robotic Technologies

Spinal robotic technologies, motorised wearable exoskeletons, are backpack-style, semi-rigid thoracolumbar supports that reduce back pain and stress, improve posture, and facilitate early mobility in spinal osteoporotic fractures. Passive Soft Exo-Suits are lightweight devices made of flexible materials that store and release energy during movement, such as the Personal Lift Augmentation Device (PLAD) and Smart

Suit Lite (SSL). Passive Rigid Exoskeletons are devices that use rigid structures and mechanical components, such as springs or beams, to provide support for the spine, for example, Bending Non-Demand Return (BNDR), LAEVO, and SPEXOR. Active Robotic Orthoses, like Back-Support Muscle Suits, use motorised actuators to actively enhance or power the joints. The use of these technologies for OSF has some limitations, such as device weight, comfort and wearability, and psychological and technical barriers (Mak & Accoto, 2021).

6.4.8 Other Emerging Technologies

Low-energy Extracorporeal Shockwave Therapy (ESWT) shows potential for improving osteoporotic fracture healing. Experimental studies demonstrate this therapy as a main or adjunctive treatment, which can enhance both qualitative and quantitative aspects of osteoporotic fracture healing, offering a simple and cost-efficient option (Mak & Accoto, 2021). Pulsed electromagnetic field therapy has also demonstrated positive effects on pain reduction, body function, Quality of Life (QoL), and bone loss prevention in PMW after spinal fracture surgery (Byalovsky et al., 2021).

6.5 Emerging Technologies in the Rehabilitation of Breast Cancer-Related Lymphedema Disability

BCRL or post-mastectomy lymphedema is a chronic, progressive PD that develops after oncological and surgical treatments affecting the lymphatic system. Conventional interventions are supervised therapeutic exercises, manual lymphatic drainage, or compression therapy. These provide episodic clinical care, are resource-intensive, and difficult to sustain over the long term, whereas technological interventions enable continuous and objective surveillance of lymphedema-related disability, improve continuity of care by extending rehabilitation beyond clinical settings, enhance task-oriented motor recovery, and improve adherence in an engaging, safe, and feedback-driven environment.

6.5.1 Virtual Reality-Based Rehabilitation for BCRL Disability

VR-based interventions, a technology-based exercise plan, not only train biomechanical factors linked to BCRL but also modify cognitive and contextual factors that influence disability. The commonly applied VR interventions included: Xbox

360 Kinect-based VR training (using Kinect Sports I), Vuzix Wrap 1200VR head-mounted glasses, Nintendo Wii game, the Greener Gamer's Nature Treks VR relaxation application, The BrightArm Duo Rehabilitation System, and CD–ROM-based scenarios. The recommended dosage for VR-based therapy is reported to be 6–8 weeks, of 20–50 min for 2 days per week. VR rehabilitation systems have been shown to benefit arm pain, shoulder ROM, muscle strength, handgrip strength, arm volume, upper-limb functionality, recovery, and compliance. Real-time feedback, an engaging environment, and gamified exercise tasks increase women's motivation and reduce the fear of movement that can exacerbate lymphedema (Bu et al., 2022). So, VR should be integrated into the standard post-mastectomy lymphedema management protocol, and improve shoulder mobility and pain outcomes (Bani Mohammad & Ahmad, 2025).

6.5.2 Digital Health Platforms, Mobile Applications, and Telerehabilitation

Digital health platforms, mobile applications, and telerehabilitation (TR) systems play a crucial role in managing BCRL by extending care beyond clinical settings. These technologies include mobile self-management apps, wearable devices integrated with smartphone applications, and web-based systems for monitoring symptoms, guiding exercises, and enabling consultation with the clinician.

For instance, the Lymphedema Self-Management App, a mobile app with nurse support, significantly improves adherence and self-care behaviours (Hemati et al., 2024). The TOLF System, a web- and mobile-based lymphatic exercise program, demonstrated significant reductions in pain severity and limb swelling, enhancing symptom management for women with BCRL over a 12-week program (Fu et al., 2022). Additionally, LymVol and LymphaTech Lite are mHealth applications focused on real-time limb measurement and swelling tracking. These apps support remote patient monitoring while providing healthcare providers with accurate data. TR programs, which offer remote exercise supervision, have been shown to improve upper-limb function, reduce pain, and increase handgrip strength, achieving comparable results to in-person sessions. These interventions are typically delivered over several weeks, enabling continuous monitoring and personalised rehabilitation (Tenório et al., 2025).

Digital health solutions thus offer an accessible, measurable, and patient-centred approach to managing BCRL, supporting early detection, continuous engagement, and improved functional outcomes. Their integration into post-mastectomy care can significantly enhance QoL and long-term recovery.

6.5.3 Artificial Intelligence and Wearable Technologies

AI-enabled wearable technologies play a pivotal role in BCRL rehabilitation. Ongoing objective monitoring, real-time surveillance of physiological and functional changes, remote clinical decision-making, and timely intervention help prevent the progression of post-mastectomy conditions. A wide range of AI-enabled technologies is currently used, including wearable motion and swelling sensors, mobile rehabilitation programs, machine-learning diagnostic models, augmented-reality feedback systems, and integrated multimodal frameworks combining wearables with virtual rehabilitation tools. Wearable devices equipped with AI analytics help identify upper-limb movement patterns and oedema-related changes. These devices generate automated alerts for therapists and improve follow-up care. In rehabilitation programs, remote supervision with adaptive intervention is also available, with some systems utilising lightweight AI algorithms embedded in sensor devices alongside mobile applications. Early and pre-clinical detection or prediction of lymphedema onset is achieved by diagnostic platforms trained on sensor-derived data. These platforms further enhance the classification accuracy. Additionally, the machine-learning sensors and augmented-reality platforms provide personalised, real-time feedback, promoting a standardised yet personal rehabilitation plan (Fu et al., 2018).

These technologies enhance exercise adherence, objectively quantify functional arm use, support ADLS, reduce the risk of complications, and improve QoL. The arm-movement sensors can objectively assess the real-time performance of the upper limb for 24 h. Intervention protocols commonly involve daily wear and continuous monitoring over multi-week periods, enabling longitudinal tracking of recovery and personalised adjustments to therapy (Eldaly et al., 2022). Collectively, these findings indicate that AI-driven wearable technologies provide continuous, personalised, and objective surveillance of disability progression, facilitating earlier intervention and improved long-term outcomes. Therefore, integration of these systems into routine post-mastectomy care pathways is strongly recommended to enhance early detection, adherence, functional recovery, and survivorship management in women with BCRL.

6.5.4 Robotic and Exoskeleton-Assisted Upper-Limb Rehabilitation

Robotic and exoskeleton-assisted technologies are increasingly utilised to support upper-limb rehabilitation following breast cancer treatment. These technologies aid recovery when traditional therapy options are limited because of pain, weakness, fatigue, or apprehension about moving an affected limb. These technology-driven systems reduce therapists' burden and restore function by promoting motor relearning through intensive, task-specific, and precisely controlled repetitions. Powered robotic exo-support devices and wearable upper-limb exoskeletons use sensors, actuators, and programmable control algorithms to control and guide the multi-joint movement of the shoulders, elbows, and wrists to move along defined biomechanical trajectories. They use real-time measurements of the patient's effort to determine and adjust the amount of assistance that the patient receives (Atiaa et al., 2025).

Commonly reported platforms include therapeutic robotic systems for oncological rehabilitation and wearable multi-degree-of-freedom exoskeletons such as the CLEVERarm. These therapeutic devices and exoskeletons provide real-time kinematic feedback to the patient, which often alters the patient's neuromuscular system's focus through adjustable resistance to the patient's movements (Zeiaee et al., 2021). Many studies show that robotic-assisted physical therapy improves range of motion and reduces fatigue in women with lymphedema, making it a valuable option in the community oncology setting. Additionally, when robotic and exoskeleton technologies are coupled with virtual or sensor-based feedback systems, patient motivation, adherence to therapy, and functional performance are improved. These interventions are typically delivered in supervised sessions multiple times per week over several weeks (approximately 6–10 weeks). Eventually, a gradual reduction in mechanical assistance is implemented as the patient's voluntary control improves (Dandoulakis, 2025).

In post-reconstruction populations, robotic systems powered by AI and integrated with wearable feedback devices may help reduce complications and support recovery. Given the potential benefits of robotic and exoskeleton technologies for safe, high-intensity, and customised rehabilitation, their use in contemporary post-mastectomy rehabilitation is justified, particularly for improving upper-limb mobility, muscular strength, and functional independence.

6.5.5 *Smart Textiles and Emerging Compression Technologies*

Smart textiles and emerging compression technologies are transforming BCRL management by offering continuous, adaptive, and real-time compression therapy. These technologies, including 4D-printed polymer nanocomposite garments and sensor-based lingerie, provide dynamic compression that automatically adjusts to swelling levels, improving both comfort and treatment efficacy. The 4D-printed garments are equipped with embedded sensors that monitor limb volume and adjust pressure accordingly, enhancing symptom management by ensuring optimal compression. Additionally, smart lingerie designed for post-mastectomy recovery supports psychological healing by offering adaptive compression while also providing real-time monitoring of oedema. These garments help reduce discomfort and improve the consistency of compression therapy, leading to better patient adherence (Cheng et al., 2025).

Emerging compression technologies, such as wearable compression garments with built-in sensors, like Compression Sleeves, and Intermittent Pneumatic Compression devices, enable more precise and individualised care by adjusting compression based on real-time data from the patient's condition. These advancements enable healthcare providers to offer a more personalised approach to lymphedema management, promoting long-term recovery and enhancing patients' QoL. The integration of these smart textiles and compression devices into post-mastectomy rehabilitation is critical for optimising outcomes and providing an effective solution for BCRL management (Dandoulakis, 2025).

6.5.6 Multimodal and Technology-Integrated Ecosystems

Multimodal and integrated technology ecosystems represent the next cutting-edge option in BCRL management. These ecosystems work by combining diverse technologies to support holistic, adaptive, and patient-centred care. These ecosystems connect digital health platforms, wearable sensors, smart compression garments, AI analytics, and TR services into a single care pathway. These systems support prevention, early detection, real-time monitoring, and long-term rehabilitation.

A key emerging aspect is the integration of predictive analytics frameworks that use multi-sensor data (e.g., limb volume changes, activity levels, and physiological signals) to predict risk of lymphedema flare-ups before they manifest clinically, enabling preemptive adjustments to therapy and compression dosing. Another notable innovation is patient engagement ecosystems. These systems combine SMS/chatbot support, interactive educational modules, and automated reminders with wearable feedback loops. These mechanisms sustain motivation and self-management over prolonged periods. Some other platforms, such as cross-platform interoperability standards, allow data from robotics, mobile apps, and smart textiles to be visualised in clinician dashboards, reducing segregated care and improving decision-making. Moreover, cloud-based repositories and secure telemonitoring facilitate asynchronous clinician review, supporting remote supervision without increasing workload. Together, these ecosystem approaches emphasise continuity, personalisation, and proactive intervention, shifting BCRL rehabilitation toward a data-driven model that enhances resilience, QoL, and long-term functional outcomes (Apkhanova et al., 2024).

6.6 Emerging Technologies in the Management of Post-hysterectomy Chronic Pelvic Pain

The rehabilitation plan for persistent structural and functional impairments and disability evolving in women having continuous CPP after undergoing hysterectomy is guided by the underlying mechanisms and pathophysiology of the medical condition. Contemporary rehabilitation approaches, such as pelvic floor muscle training, pain modulation with physiotherapeutic agents, massage, myofascial release, trigger-point compression, and self-management, are effective for reducing pain and improving function. Yet the role of emerging technologies for his multifactorial condition, in rehabilitating structural impairments, neuromuscular dysfunctions, and the central sensitisation process, is evolving as promising adjuncts to reduce the burden of disability (Fuentes-Márquez et al., 2019; Starzec-Proserpio et al., 2025).

6.6.1 Virtual Reality (VR) and Immersive Digital Therapeutics

VR platforms are utilised to rehabilitate a variety of health conditions leading to physical disabilities. It simulates real-life situations, functional tasks, and movements and engages users in artificial computer-generated environments. VR stimulates neural circuits through multiple sensory feedback channels and promotes patient learning and recovery during the disability process. Within VR applications, there are two distinct media: immersive and non-immersive. The main idea of VR-based chronic pain management is to create an efficient distraction from painful stimuli compared to traditional methods. VR application modifies pain perceptions, increases pain tolerance and thresholds. A recent systematic review (SR) has found a positive association of VR interventions with chronic pain outcome measures, including intensity, functional capacity, disability, mobility, psychological status, and QoL (Goudman et al., 2022).

Despite the proven efficacy of VR-based interventions on chronic pain mechanisms, only a handful of clinical trials have been conducted on pelvic floor rehabilitation using VR as an intervention for chronic pain experienced by women after hysterectomy. A pilot study tested two types of VR interventions on a group of women with CPP due to endometriosis. After 14 weeks of intervention, VR-based relaxation therapy showed more consistent results in pain, stress, and depression scores reduction as compared to the activity-based app. Yet confirmation through large trials is suggested (Pakebusch et al., 2025). Immersive VR (IVR) is believed to be an effective and low-risk therapy for CPP. A recent SR & Meta-analysis reviewed 3 RCTs with $n = 169$ endometriosis-based CPP women who used immersive VR and showed a moderate reduction in pain compared to sham (González-Lara et al., 2025). A study explored the comparative immediate effects of a single session of self-managed VR versus supervised telehealth exercises and found both equally favourable for women with pelvic pain (Lutfi et al., 2023).

6.6.2 Digital Pelvic Rehabilitation Platforms

Using the internet or E-health systems is a highly effective way to educate patients about various conditions. App-Mohedo®, the first user-friendly, multiplatform, multilingual mobile app that uses graded motor imagery and left–right judgement tasks, was developed to augment conventional management of CPP. It included almost 550 pelvic floor images, assembled into three difficulty levels (Díaz-Mohedo et al., 2024). Another pilot project developed and tested a smartphone-based CBT program through an application named appease, which delivered a 28-day intervention to women with CPP by clinicians, including patient education, cognitive restructuring, relaxation, and mindfulness. 75% of women completed the study and found it helpful, easy to use, engaging, and effective in reducing pain catastrophising and increasing self-efficacy.

6.6.3 Sensor-Augmented Biofeedback Systems

BF is an instrument-assisted training used for the rehabilitation of pelvic floor dysfunctions, by using either electromyography or pressure sensors. As dysfunctional PFMs are associated with PH-CPP, BF training is an effective therapeutic option for this condition, as it helps patients understand the exact contraction or relaxation of the pelvic floor muscles using verbal, visual, or acoustic feedback (Calderone et al., 2025; Wagner et al., 2022). EMG BF is used for dual purposes of diagnosis and intervention, using either surface or vaginal electrodes, to measure the electrical activity of PFMs and offers real-time monitoring and guided exercises. It enhances self-control over psychophysiological processes in chronic and complex pain (Calderone et al., 2025). The EAU's 2019 guidelines state that CPP is an effective adjuvant therapy option for myofascial therapy in hypertonic pelvic floor dysfunctions, such as in women who have undergone hysterectomy (Engeler et al., 2019). An evaluation of clinical practice guidelines indicates the highest level of recommendations for BF training as an adjunct to exercises for overactive PFMs (Ghai et al., 2021).

A recent SR, analysing 37 studies using BF for various pelvic pain conditions, anorectal and urogynaecological disorders, demonstrates tentative evidence that BF-assisted training produces preliminary positive effects on pelvic pain, associated symptoms, and patients' QoL (Wagner et al., 2022). Another prospective RCT proved the effectiveness of a multimodal rehab protocol, combining BF and capacitive–resistive radiofrequency therapy, in reducing pain and improving functional improvement in women with CPP and dyspareunia (Fernández-Cuadros et al., 2020). The advent of wireless intravaginal BF devices brings together cutting-edge technologies: real-time feedback, smartphone connectivity, and gamification. Such devices empower women with disabilities due to chronic pain through efficient user engagement and adherence to rehabilitation protocols (Tank & Rathod, 2025).

6.6.4 *Neuromodulation and Pain Modulation Technologies*

Neuromodulation, an emerging treatment based on electrical stimulation, is used for various chronic pain conditions. Since ancient times, electrotherapeutic neuromodulation has been practised for pain management. Since 1960, deep-brain and spinal cord stimulation have been used to treat intractable pain. Gradually, technological advances in neuromodulation have yielded significant improvements in patient outcomes. In conditions such as CPP, neuromodulation is often considered the last bastion of potentially effective therapy, but its efficacy depends on the placement of leads to target specific areas (Hunter et al., 2013; Martellucci et al., 2012).

The use and efficacy of an array of neuromodulatory modalities, including "Sacral nerve, pudendal nerve, percutaneous tibial nerve stimulation, dorsal root ganglion stimulation, dorsal column neuromodulation", have been reviewed recently (Hao et al., 2022). The potential positive and expanding role of various etiologies of CPP has been demonstrated, and the need for rigorous, well-designed, large, multi-center clinical trials to demonstrate efficacy in patient pain, functionality, and QoL has been highlighted. Below are mentioned the mechanisms of some emerging neuromodulating technologies (Gish et al., 2024).

Sacral nerve roots Neuromodulation modulate the sensory pathways for the pelvic floor (S2–S4 nerve roots) to treat CPP and other related conditions. The unilateral S3 nerve root is mostly targeted in reported clinical cases. It is a more selective and stable pathway for delivering electrical pulses that control sensation and function in the pelvis, aiming to dampen pain in the pelvic region and adjacent abdominal areas. This pathway has anatomical advantages, as sacral neuroanatomy is less packed with fibres, less mobile, and has a thinner dorsal cerebrospinal fluid layer than the conus medullaris and cauda equina. The leads are inserted via different surgical approaches (Lavonius et al., 2017; Marcelissen et al., 2010; Sokal et al., 2015).

Dorsal Root Ganglion Stimulation (DRGS) modulates nociceptive transmission, reducing action potential propagation and neuronal excitability. To stimulate the primary sensory neurons responsible for pain transmission, the dorsal root ganglion, which contains the cell bodies of primary sensory neurons, is stimulated in the lateral epidural space, from the lower thorax to the sacral vertebrae. The evidence for this stimulation is very low compared to traditional stimulation for pain and dysfunction (Nagpal et al., 2021).

Dorsal Column Spinal Cord Stimulation (DCSCS) modulates ascending neuronal tracts and addresses hypersensitivity. To stimulate ascending nerve tracts of the spinal cord, the leads are placed in the epidural space from the mid to lower thorax (Hao et al., 2022).

Pudendal nerve stimulation (PNS) induces selective and targeted peripheral analgesia. The nerve is stimulated along its course using a minimally invasive needle approach with neurophysiological guidance. The majority of studies have used the posterior-ischial-rectal approach to place the quadripolar leads (Peters et al., 2007).

Transcranial direct current stimulation (tDCS) also modulates CPP by central nervous system pain processing; studies have shown significant improvements in subjective and objective pain outcome measures compared to sham. A recent clinical trial in women with endometriosis-associated CPP used an anodal electrode on the primary motor cortex for 10 days. It showed a reduction in pain score and a sustained increase in pain pressure threshold after 1 week of intervention. Strong clinical trials are further recommended (Mechsner et al., 2023).

Transcutaneous Electrical Nerve Stimulation (TENS) provides segmental pain blockade through gait-control theory. The high-frequency mechanical stimulation activates large diameter fibres while inhibiting the nociceptive small unmyelinated nerve fibres at the dorsal horn of the spinal cord. High-intensity and high-frequency TENS application also engages descending inhibitory pathways, as endogenous opioids (GABA and Serotonin) are released to reduce central sensitisation and hyperalgesia. Recent SRs and meta-analyses suggest that high-quality, methodologically rigorous, large-sample studies are needed to establish the effects of TENS on the chronicity of pain in women after hysterectomy (Mendes et al., 2025).

Percutaneous Tibial Nerve Stimulation (PTNS) is a minimally invasive procedure that mitigates CPP by neuromodulating the sacral reflex arc via stimulation of the posterior tibial nerve, which shares sensory fibres with the sacral nerve roots (S2-S4) involved in CPP processing. It modulates visceral hypersensitivity and central sensitisation following hysterectomy (Istek et al., 2014).

6.7　Conclusion and Future Directions

Physical disabilities related to women's health conditions are the complex interplay of biological and sociocultural determinants. Some multidimensional disability conditions, such as PMOP and related fragility fractures, post-mastectomy lymphedema, and post-hysterectomy CPP, are not merely isolated clinical conditions. These affect a woman at all levels of disability spectrum, such as loss of body structure or function, activity limitation, and participation restrictions. Age-linked hormonal transitions, reproductive surgeries, chronic pain, and central sensitisation are the key disability drivers in these women's health conditions.

The integration of technologies into rehabilitation protocols is indeed a paradigm shift, facilitating individualised, data-informed clinical decisions. AI-based risk prediction models, VR-based exercise protocols, balance platforms, wearable sensors for postural correction and TR, and other technologies can support osteoporotic fracture prevention and bone density optimisation. In BCRL, upper-limb function can be restored through early detection and precision therapy enabled by AI-enabled monitoring, smart compression textiles, and robotic-assisted upper-limb rehabilitation. Following hysterectomy, chronic pelvic, peripheral, and central pain mechanisms, as well as neuromuscular dysfunction, can be managed with sensor-augmented BF, digital health platforms, and neuromodulation technologies. These technologies are useful for augmenting disability assessment, evaluation, continuous monitoring, therapeutic intervention delivery, and adherence to the protocol. In women's health, such innovative solutions are more relevant, as their disabilities are often chronic, stigmatised, and compromised by limited access to healthcare and their caregiver role.

However, despite the advancement in technology, the limitations of digital literacy gaps in women, sociocultural stigma, and health system disparities, make the implementation uneven. This signifies the need for policies that accommodate and prioritise gender-sensitive design, affordability, and inclusivity. With the integration of innovative technologies, the reactive disability management among women can be shifted to proactive functional restoration. There should be strong, ongoing collaboration among clinicians, rehabilitation scientists, patient communities, biomedical engineers, policymakers, neuroscientists, psychologists, and AI and information technology experts to deliver more advanced, smart, flexible, and cost-effective solutions. The long-term effectiveness of these technologies in disability management should be tested through rigorous clinical trials that recruit a diverse population. Such revolutions in disability management can translate into steady improvements in women's independence, participation, and QoL worldwide.

References

Ali, A., Huszti, E., Noordin, S., Ali, U., & Sale, J. E. M. (2024). Examining treatment targets and equity in bone-active medication use within secondary fracture prevention: A systematic review and meta-analysis. *Osteoporosis International: A Journal Established as Result of Cooperation between the European Foundation for Osteoporosis and the National Osteoporosis Foundation of the USA, 35*(9), 1497–1511. https://doi.org/10.1007/s00198-024-07078-5

Apkhanova, T. V., Валерьевна, А. Т., Konchugova, T. V., Венедиктовна, К. Т., Kulchitskaya, D. В., Борисовна, К. Д., Yurova, O. V., Валентиновна, Ю. О., Styazhkina, E. M., Михайловна, С. Е., Marfina, T. V., Владимировна, М. Т., Agasarov, L. G., Георгиевич, А. Л., Vasileva, V. A., Александровна, В. В., Berezkina, E. S., & Сергеевна, Б. Е. (2024). New non-drug technologies for lymphedema associated with breast cancer: A review. *Bulletin of Rehabilitation Medicine, 23*(3), 40–51. https://doi.org/10.38025/2078-1962-2024-23-3-40-51

Atiaa, A. G., Mostafa, M. M., & Ellakwa, D. E.-S. (2025). Technology-enhanced compression and AI-integrated lymphedema care: A narrative review. *Irish Journal of Medical Science (1971–), 194*(6), 1957–1976. https://doi.org/10.1007/s11845-025-04101-4

Bakar, Y. (2025). Breast cancer-related lymphedema: From description to management. In *Managing side effects of breast cancer treatment* (pp. 99–114). Springer.

Bani Mohammad, I., & Ahmad, M. (2025). Using virtual therapy for lymphedema and disability post-mastectomy: Meta-analysis with systematic review. *Breast Care, 20*(6), 437–451.

Beaudoin, C., Moore, L., Gagné, M., Bessette, L., Ste-Marie, L. G., Brown, J. P., & Jean, S. (2019). Performance of predictive tools to identify individuals at risk of non-traumatic fracture: A systematic review, meta-analysis, and meta-regression. *Osteoporosis International, 30*(4), 721–740. https://doi.org/10.1007/s00198-019-04919-6

Beck, A. C., & Lizarraga, I. M. (2024). Long-term burden of breast cancer-related lymphedema. *Current Breast Cancer Reports, 16*(2), 251–259.

Bethi, S. R., RajKumar, A., Vulpi, F., Raghavan, P., & Kapila, V. (2020). *Wearable inertial sensors for exergames and rehabilitation.* In 2020 42nd Annual International Conference of the IEEE Engineering in Medicine & Biology Society (EMBC), pp. 4579–4582. https://doi.org/10.1109/EMBC44109.2020.9175428

Bilek, L. D., Flores, L. E., Waltman, N., Mack, L. R., Smith, K., Hillstrom, D., Griffin, M., Yecies, L., & Jaasma, M. (2023). FRI681 Osteoboost™ is effective in preserving bone strength and density of the spine in women with low bone mass. *Journal of the Endocrine Society, 7*(Suppl. 1), bvad114.449. https://doi.org/10.1210/jendso/bvad114.449

Biz, C., Khamisy-Farah, R., Puce, L., Szarpak, L., Converti, M., Ceylan, H. İ, Crimì, A., Bragazzi, N. L., & Ruggieri, P. (2024). Investigating and practicing orthopedics at the intersection of sex and gender: Understanding the physiological basis, pathology, and treatment response of orthopedic conditions by adopting a gender lens: A narrative overview. *Biomedicines, 12*(5), 974. https://doi.org/10.3390/biomedicines12050974

Bu, X., Ng, P. H., Xu, W., Cheng, Q., Chen, P. Q., Cheng, A. S., & Liu, X. (2022). The effectiveness of virtual reality–based interventions in rehabilitation management of breast cancer survivors: Systematic review and meta-analysis. *JMIR Serious Games, 10*(1), e31395.

Byalovsky, Y. Y., Ivanov, A. V., & Rakitina, I. S. (2021). Effects of a pulsed electromagnetic field on the course of osteoporosis in postmenopausal women. *Russian Journal of Physiotherapy, Balneology and Rehabilitation, 20*(5), 385–395.

Byun, H. K., Chang, J. S., Im, S. H., Kirova, Y. M., Arsene-Henry, A., Choi, S. H., Cho, Y. U., Park, H. S., Kim, J. Y., Suh, C.-O., Keum, K. C., Sohn, J. H., Kim, G. M., Lee, I. J., Kim, J. W., & Kim, Y. B. (2021). Risk of lymphedema following contemporary treatment for breast cancer: An analysis of 7617 consecutive patients from a multidisciplinary perspective. *Annals of Surgery, 274*(1), 170–178. https://doi.org/10.1097/SLA.0000000000003491

Calderone, A., Masi, V. M. M., Luca, R. D., Gangemi, A., Bonanno, M., Floridia, D., Corallo, F., Morone, G., Quartarone, A., Maggio, M. G., & Calabrò, R. S. (2025). The impact of biofeedback in enhancing chronic pain rehabilitation: A systematic review of mechanisms and outcomes. *Heliyon, 11*(2). https://doi.org/10.1016/j.heliyon.2025.e41917

Calikyan, A., Silverberg, J., & McLeod, K. M. (2023). Osteoporosis screening disparities among ethnic and racial minorities: A systematic review. *Journal of Osteoporosis, 2023*, 1277319. https://doi.org/10.1155/2023/1277319

Chandran, M., Thiyagarajan, J. A., Alokail, M., Bruyère, O., Harvey, N. C., Rizzoli, R., Veronese, N., & Reginster, J.-Y. (2026). WHO benchmarks for equitable hip-fracture care and osteoporosis treatment in older people. *Nature Reviews. Rheumatology, 22*(1), 62–70. https://doi.org/10.1038/s41584-025-01319-5

Cheng, H., Gong, J., Yu, L., Feng, X., Ou, W., Wang, H., Zhong, H., & Zhang, E. M. (2025). Compression sleeves for prevention and treatment of breast cancer-related lymphedema: A systematic review and meta-analysis. *Breast Cancer Research and Treatment, 215*(1), 41. https://doi.org/10.1007/s10549-025-07846-9

Cheong, Y. C., Smotra, G., & Williams, A. C. de C. (2014). Non-surgical interventions for the management of chronic pelvic pain. *The Cochrane Database of Systematic Reviews, 2014*(3), CD008797. https://doi.org/10.1002/14651858.CD008797.pub2

CORE: A global aggregation service for open access papers. (n.d.). *ResearchGate.* https://doi.org/10.1038/s41597-023-02208-w

Cortés-Pérez, I., Díaz-Fernández, Á., Osuna-Pérez, M. C., García-López, H., Romero-Del-Rey, R., & Obrero-Gaitán, E. (2025). Virtual reality-based therapy improves balance, quality of life, and mitigates pain and fear of falling in women with bone mineral density loss: A meta-analysis of randomized controlled trials. *Life, 15*(11), 1654. https://doi.org/10.3390/life15111654

Cultrera, P., Pratelli, E., Petrai, V., Postiglione, M., Zambelan, G., & Pasquetti, P. (2010). Evaluation with stabilometric platform of balance disorders in osteoporosis patients. A proposal for a diagnostic protocol. *Clinical Cases in Mineral and Bone Metabolism: The Official Journal of the Italian Society of Osteoporosis, Mineral Metabolism, and Skeletal Diseases, 7*(2), 123–125.

Dabas, D., & Singh, K. N. (2024). Understanding osteoporosis knowledge, beliefs, and behaviors in the older population: A literature review. *Journal of Nursing Science and Professional Practice, 1*(2), 51. https://doi.org/10.4103/JNSPP.JNSPP_8_24

Dance, A. (2019). Why the sexes don't feel pain the same way. *Nature, 567*(7749), 448–450. https://doi.org/10.1038/d41586-019-00895-3

Dandoulakis, E. (2025). *Innovative strategies in lymphedema management post-breast reconstruction: A comprehensive systematic overview* (SSRN Scholarly Paper No. 5678094). Social Science Research Network. https://doi.org/10.2139/ssrn.5678094

des Bordes, J., Prasad, S., Pratt, G., Suarez-Almazor, M. E., & Lopez-Olivo, M. A. (2020). Knowledge, beliefs, and concerns about bone health from a systematic review and metasynthesis of qualitative studies. *PloS One, 15*(1), e0227765. https://doi.org/10.1371/journal.pone.0227765

Díaz-Mohedo, E., Carrillo-León, A. L., Calvache-Mateo, A., Ptak, M., Romero-Franco, N., & Carlos-Fernández, J. (2024). App-Mohedo®: A mobile app for the management of chronic pelvic pain. A design and development study. *International Journal of Medical Informatics, 186*, 105410. https://doi.org/10.1016/j.ijmedinf.2024.105410

DiSipio, T., Rye, S., Newman, B., & Hayes, S. (2013). Incidence of unilateral arm lymphoedema after breast cancer: A systematic review and meta-analysis. *The Lancet Oncology, 14*(6), 500–515.

Eldaly, A. S., Avila, F. R., Torres-Guzman, R. A., Maita, K., Garcia, J. P., Serrano, L. P., & Forte, A. J. (2022). Artificial intelligence and lymphedema: State of the art. *Journal of Clinical and Translational Research, 8*(3), 234.

Engeler, D., Bananowski, A., Berghmans, B., Borovicka, J., Cottrell, A., Elneil, S., Hughes, J., Messelink, E., & Williams, A. (2019). *EAU guidelines on chronic pelvic pain.*

Fernández-Cuadros, M., Kazlauskas, S., Albaladejo-Florin, M., Robles-López, M., Laborda-Delgado, A., De La Cal-Alvarez, C., & Pérez-Moro, O. (2020). Effectiveness of multimodal rehabilitation (biofeedback plus capacitive-resistive radiofrequency) on chronic pelvic pain and dyspareunia: Prospective study and literature review. *Rehabilitación, 54*(3), 154–161.

Fu, M. R., Axelrod, D., Guth, A. A., Scagliola, J., Rampertaap, K., El-Shammaa, N., Qiu, J. M., McTernan, M. L., Frye, L., Park, C. S., Yu, G., Tilley, C., & Wang, Y. (2022). A web- and mobile-based intervention for women treated for breast cancer to manage chronic pain and symptoms related to lymphedema: Results of a randomized clinical trial. *JMIR Cancer, 8*(1), e29485. https://doi.org/10.2196/29485

Fu, M. R., Wang, Y., Li, C., Qiu, Z., Axelrod, D., Guth, A. A., Scagliola, J., Conley, Y., Aouizerat, B. E., & Qiu, J. M. (2018). Machine learning for detection of lymphedema among breast cancer survivors. *Mhealth, 4*, 17.

Fuentes-Márquez, P., Cabrera-Martos, I., & Valenza, M. C. (2019). Physiotherapy interventions for patients with chronic pelvic pain: A systematic review of the literature. *Physiotherapy Theory and Practice, 35*(12), 1131–1138. https://doi.org/10.1080/09593985.2018.1472687

Ghai, V., Subramanian, V., Jan, H., Loganathan, J., Doumouchtsis, S. K., & CHORUS: An International Collaboration for Harmonising Outcomes, Research and Standards in Urogynaecology and Women's Health (i-chorus.org). (2021). Evaluation of clinical practice guidelines (CPG) on the management of female chronic pelvic pain (CPP) using the AGREE II instrument. *International Urogynecology Journal, 32*(11), 2899–2912. https://doi.org/10.1007/s00192-021-048 48-1

Gish, B., Langford, B., Sobey, C., Singh, C., Abdullah, N., Walker, J., Gray, H., Hagedorn, J., Ghosh, P., & Patel, K. (2024). Neuromodulation for the management of chronic pelvic pain syndromes: A systematic review. *Pain Practice, 24*(2), 321–340.

González-Lara, C., Obrero-Gaitán, E., Piñar-Lara, M., Desdentado-Guillén, J. M., Peinado-Rubia, A. B., & Cortés-Pérez, I. (2025). Immersive virtual reality interventions for endometriosis-related pelvic pain: Systematic review with meta-analysis. *Journal of Minimally Invasive Gynecology, 32*(11), 960–969. https://doi.org/10.1016/j.jmig.2025.06.017

Goudman, L., Jansen, J., Billot, M., Vets, N., De Smedt, A., Roulaud, M., Rigoard, P., & Moens, M. (2022). Virtual reality applications in chronic pain management: Systematic review and meta-analysis. *JMIR Serious Games, 10*(2), e34402.

Greenspan, J. D., Craft, R. M., LeResche, L., Arendt-Nielsen, L., Berkley, K. J., Fillingim, R. B., Gold, M. S., Holdcroft, A., Lautenbacher, S., Mayer, E. A., Mogil, J. S., Murphy, A. Z., Traub, R. J., & Consensus Working Group of the Sex, Gender, and Pain SIG of the IASP. (2007). Studying sex and gender differences in pain and analgesia: A consensus report. *Pain, 132*(Suppl. 1), S26–S45. https://doi.org/10.1016/j.pain.2007.10.014

Gupta, A., Maslen, C., Vindlacheruvu, M., Abel, R. L., Bhattacharya, P., Bromiley, P. A., Clark, E. M., Compston, J. E., Crabtree, N., Gregory, J. S., Kariki, E. P., Harvey, N. C., McCloskey, E., Ward, K. A., & Poole, K. E. S. (2022). Digital health interventions for osteoporosis and post-fragility fracture care. *Therapeutic Advances in Musculoskeletal Disease, 14.* https://doi.org/10.1177/1759720X221083523

Hao, D., Yurter, A., Chu, R., Salisu-Orhurhu, M., Onyeaka, H., Hagedorn, J., Patel, K., D'Souza, R., Moeschler, S., & Kaye, A. D. (2022). Neuromodulation for management of chronic pelvic pain: A comprehensive review. *Pain and Therapy, 11*(4), 1137–1177.

Hemati, M., Rivaz, M., & Khademian, Z. (2024). Lymphedema self-management mobile application with nurse support for post breast cancer surgery survivors: Description of the design process and prototype evaluation. *BMC Cancer, 24,* 973. https://doi.org/10.1186/s12885-024-12744-2

Hewage, K., Fosker, S., Leckie, T., Venn, R., Gonçalves, A.-C., Koulouglioti, C., & Hodgson, L. E. (2023). The Hospital to Home study (H2H): Smartwatch technology-enabled rehabilitation following hip fracture in older adults, a feasibility non-randomised trial. *Future Healthcare Journal, 10*(1), 14–20. https://doi.org/10.7861/fhj.2022-0101

Hunter, C., Davé, N., Diwan, S., & Deer, T. (2013). Neuromodulation of pelvic visceral pain: Review of the literature and case series of potential novel targets for treatment. *Pain Practice, 13*(1), 3–17.

Istek, A., Gungor Ugurlucan, F., Yasa, C., Gokyildiz, S., & Yalcin, O. (2014). Randomized trial of long-term effects of percutaneous tibial nerve stimulation on chronic pelvic pain. *Archives of Gynecology and Obstetrics, 290*(2), 291–298. https://doi.org/10.1007/s00404-014-3190-z

Ji, M.-X., & Yu, Q. (2015). Primary osteoporosis in postmenopausal women. *Chronic Diseases and Translational Medicine, 1*(1), 9–13. https://doi.org/10.1016/j.cdtm.2015.02.006

Johnell, O., & Kanis, J. (2006). An estimate of the worldwide prevalence and disability associated with osteoporotic fractures. *Osteoporosis International, 17*(12), 1726–1733.

Jordan, K., & Cooper, C. (2002). Epidemiology of osteoporosis. *Best Practice & Research. Clinical Rheumatology, 16*(5), 795–806.

Jouanny, C., Abhyankar, P., & Maxwell, M. (2024). A mixed methods systematic literature review of barriers and facilitators to help-seeking among women with stigmatised pelvic health symptoms. *BMC Women's Health, 24*(1), 217. https://doi.org/10.1186/s12905-024-03063-6

Jung, J., Yu, J., & Kang, H. (2012). Effects of virtual reality treadmill training on balance and balance self-efficacy in stroke patients with a history of falling. *Journal of Physical Therapy Science, 24*(11), 1133–1136.

Kehlet, H., Jensen, T. S., & Woolf, C. J. (2006). Persistent postsurgical pain: Risk factors and prevention. *Lancet, 367*(9522), 1618–1625. https://doi.org/10.1016/S0140-6736(06)68700-X

Kemmak, A. R., Rezapour, A., Jahangiri, R., Nikjoo, S., Farabi, H., & Soleimanpour, S. (2020). Economic burden of osteoporosis in the world: A systematic review. *Medical Journal of the Islamic Republic of Iran, 34,* 154.

Kovac, S. R. (2014). Route of hysterectomy: An evidence-based approach. *Clinical Obstetrics and Gynecology, 57*(1), 58–71.

Lamvu, G., Carrillo, J., Ouyang, C., & Rapkin, A. (2021). Chronic pelvic pain in women: A review. *JAMA, 325*(23), 2381–2391. https://doi.org/10.1001/jama.2021.2631

Lavonius, M., Suvitie, P., Varpe, P., & Huhtinen, H. (2017). Sacral neuromodulation: Foray into chronic pelvic pain in end stage endometriosis. *Case Reports in Neurological Medicine, 2017*(1), 2197831.

Li, J., Jiang, Y., Liu, Y., & Shao, Z. (2017). Identify high risk estrogen receptor-positive breast cancer patients for extended endocrine therapy. *The Breast, 31,* 173–180.

Li, Q., Liang, L., Gao, C., & Zong, B. (2024). Therapeutic effects of whole-body vibration on postmenopausal women with osteoporosis: A systematic review and meta-analysis. *Brazilian Journal of Medical and Biological Research, 57,* e13996.

Li, X., Chen, Z., Yue, Y., Zhou, X., Gu, S., Tao, J., Guo, H., Zhu, M., & Du, Q. (2022). Effect of wearable sensor-based exercise on musculoskeletal disorders in individuals with neurodegenerative diseases: A systematic review and meta-analysis. *Frontiers in Aging Neuroscience, 14,* 934844. https://doi.org/10.3389/fnagi.2022.934844

Liang, H., Chen, S., Shi, M., Xu, J., Zhao, C., Yang, B., Zheng, S., & Tan, J. (2025). Global epidemiology and burden of osteoporosis among postmenopausal women: Insights from the Global Burden of Disease Study 2021. *Npj Aging, 11*(1), 78.

Lorentzon, M., Johansson, H., Harvey, N. C., Liu, E., Vandenput, L., McCloskey, E. V., & Kanis, J. A. (2022). Osteoporosis and fractures in women: The burden of disease. *Climacteric: The Journal of the International Menopause Society, 25*(1), 4–10. https://doi.org/10.1080/13697137.2021.1951206

Lutfi, M., Dalleck, L. C., Drummond, C., Drummond, M., Paparella, L., Keith, C. E., Kirton, M., Falconer, L., Gebremichael, L., Phelan, C., Barry, C., Roscio, K., Lange, B., & Ramos, J. S. (2023). A single session of a digital health tool-delivered exercise intervention may provide immediate relief from pelvic pain in women with endometriosis: A pilot randomized controlled study. *International Journal of Environmental Research and Public Health, 20*(3), 1665. https://doi.org/10.3390/ijerph20031665

Madansingh, S. I., Murphree, D. H., Kaufman, K. R., & Fortune, E. (2019). Assessment of gait kinetics in post-menopausal women using tri-axial ankle accelerometers during barefoot walking. *Gait & Posture, 69,* 85–90.

Maden, T., Ergen, H. İ, Pancar, Z., Buglione, A., Padulo, J., Migliaccio, G. M., & Russo, L. (2025). The effects of virtual reality-based task-oriented movement on upper extremity function in healthy individuals: A crossover study. *Medicina, 61*(4), 668. https://doi.org/10.3390/medicina61040668

Mak, S. K. D., & Accoto, D. (2021). Review of current spinal robotic orthoses. *Healthcare, 9*(1), 70. https://doi.org/10.3390/healthcare9010070

Marcelissen, T., Van Kerrebroeck, P., & De Wachter, S. (2010). Sacral neuromodulation as a treatment for neuropathic clitoral pain after abdominal hysterectomy. *International Urogynecology Journal, 21*(10), 1305–1307.

Marchenkova, L. A., Vasilyeva, V. A., Otvetchikova, D. I., & Fesyun, A. D. (2024). Efficacy of virtual reality and mechanotherapy technologies in the rehabilitation of patients with osteoporosis after surgical treatment of femoral fractures. *Russian Journal of Geriatric Medicine* (4), 270–280. https://doi.org/10.37586/2686-8636-4-2024-270-280

Martellucci, J., Naldini, G., & Carriero, A. (2012). Sacral nerve modulation in the treatment of chronic pelvic pain. *International Journal of Colorectal Disease, 27*(7), 921–926.

Mattia, A., Hadzimustafic, N., Rivero, R., Oh, S. J., Bach, K., Brown, S., & Haykal, S. (2025). Disparities in breast cancer-related lymphedema: A systematic review of inequities and barriers in care. *Plastic and Reconstructive Surgery. Global Open, 13*(7), e6935. https://doi.org/10.1097/GOX.0000000000006935

Mechsner, S., Grünert, J., Wiese, J. J., Vormbäumen, J., Sehouli, J., Siegmund, B., Neeb, L., & Prüß, M. S. (2023). Transcranial direct current stimulation to reduce chronic pelvic pain in endometriosis: Phase II randomized controlled clinical trial. *Pain Medicine, 24*(7), 809–817. https://doi.org/10.1093/pm/pnad031

Mendes, C. F., Oliveira, L. S., Garcez, P. A., Azevedo-Santos, I. F., & DeSantana, J. M. (2025). Effect of different electric stimulation modalities on pain and functionality of patients with pelvic pain: Systematic review with META-analysis. *Pain Practice, 25*(1), e13417. https://doi.org/10.1111/papr.13417

Misal, M., Balk, E. M., Orlando, M. S., Pickett, C., Lenger, S. M., Porter, A. E., Balgobin, S., Jalloul, R. J., Encalada-Soto, D., Mamik, M. M., & Gupta, A. (2025). Predictors of persistent pain after hysterectomy for chronic pelvic pain: A systematic review. *Obstetrics and Gynecology, 146*(5), 690–699. https://doi.org/10.1097/AOG.0000000000006023

Moka, E., Aguirre, J. A., Sauter, A. R., & Lavand'homme, P. (2025). Chronic postsurgical pain and transitional pain services: A narrative review highlighting European perspectives. *Regional Anesthesia & Pain Medicine, 50*(2), 205–212. https://doi.org/10.1136/rapm-2024-105614

Nagpal, A., Clements, N., Duszynski, B., & Boies, B. (2021). The effectiveness of dorsal root ganglion neurostimulation for the treatment of chronic pelvic pain and chronic neuropathic pain of the lower extremity: A comprehensive review of the published data. *Pain Medicine, 22*(1), 49–59.

North, K., Simpson, G., Geiger, W., Cizik, A., Rothberg, D., & Hitchcock, R. (2024). Predicting the healing of lower extremity fractures using wearable ground reaction force sensors and machine learning. *Sensors, 24*(16), 5321. https://doi.org/10.3390/s24165321

Onizuka, N., & Onizuka, T. (2024). Disparities in osteoporosis prevention and care: Understanding gender, racial, and ethnic dynamics. *Current Reviews in Musculoskeletal Medicine, 17*(9), 365–372. https://doi.org/10.1007/s12178-024-09909-8

Pakebusch, V., Schlisio, B., Schönfisch, B., Brucker, S. Y., Krämer, B., & Andress, J. (2025). Virtual reality-based pain control in endometriosis: A questionnaire-based pilot study of applications for relaxation and physical activity. *Archives of Gynecology and Obstetrics, 311*(6), 1721–1731. https://doi.org/10.1007/s00404-025-08000-y

Pastore, E. A., & Katzman, W. B. (2012). Recognizing myofascial pelvic pain in the female patient with chronic pelvic pain. *Journal of Obstetric, Gynecologic & Neonatal Nursing, 41*(5), 680–691.

Perdomo, M., Davies, C., Levenhagen, K., Ryans, K., & Gilchrist, L. (2023). Patient education for breast cancer-related lymphedema: A systematic review. *Journal of Cancer Survivorship: Research and Practice, 17*(2), 384–398. https://doi.org/10.1007/s11764-022-01262-4

Peters, K. M., Feber, K. M., & Bennett, R. C. (2007). A prospective, single-blind, randomized crossover trial of sacral vs pudendal nerve stimulation for interstitial cystitis. *BJU International, 100*(4), 835–839. https://doi.org/10.1111/j.1464-410X.2007.07082.x

Raffegeau, T. E., Young, W. R., Fino, P. C., & Williams, A. M. (2023). A perspective on using virtual reality to incorporate the affective context of everyday falls into fall prevention. *JMIR Aging, 6*, e36325.

Rezaei, M. K., Torkaman, G., Bahrami, F., & Bayat, N. (2023). The effect of six week virtual reality training on the improvement of functional balance in women with type-I osteoporosis: A preliminary study. *Sport Sciences for Health, 19*(1), 185–194. https://doi.org/10.1007/s11332-022-01018-8

Riaz, S., Shakil Ur Rehman, S., Hafeez, S., & Hassan, D. (2024a). Effects of kinect-based virtual reality training on bone mineral density and fracture risk in postmenopausal women with osteopenia: A randomized controlled trial. *Scientific Reports, 14*(1), 6650. https://doi.org/10.1038/s41598-024-57358-7

Riaz, S., Shakil Ur Rehman, S., Hassan, D., & Hafeez, S. (2024b). Gamified exercise with kinect: Can kinect-based virtual reality training improve physical performance and quality of life in postmenopausal women with osteopenia? A randomized controlled trial. *Sensors, 24*(11), 3577. https://doi.org/10.3390/s24113577

Ruiz-Esteves, K. N., Teysir, J., Schatoff, D., Yu, E. W., & Burnett-Bowie, S.-A.M. (2022). Disparities in osteoporosis care among postmenopausal women in the United States. *Maturitas, 156*, 25–29. https://doi.org/10.1016/j.maturitas.2021.10.010

Sambrook, P., & Cooper, C. (2006). Osteoporosis. *The Lancet, 367*(9527), 2010–2018. https://doi.org/10.1016/S0140-6736(06)68891-0

Singh, A., & Varma, A. R. (n.d.). Whole-body vibration therapy as a modality for treatment of senile and postmenopausal osteoporosis: A review article. *Cureus, 15*(1), e33690. https://doi.org/10.7759/cureus.33690

Snell, A., Lobaina, D., Densley, S., Moothedan, E., Baker, J., Al Abdul Razzak, L., Garcia, A., Skibba, S., Dunn, A., & Follin, T. (2025). Disparities in postoperative pain management: A scoping review of prescription practices and social determinants of health. *Pharmacy, 13*(2), 34.

Sokal, P., Zieliński, P., & Harat, M. (2015). Sacral roots stimulation in chronic pelvic pain. *Neurologia i Neurochirurgia Polska, 49*(5), 307–312.

Sorge, R. E., Mapplebeck, J. C. S., Rosen, S., Beggs, S., Taves, S., Alexander, J. K., Martin, L. J., Austin, J.-S., Sotocinal, S. G., Chen, D., Yang, M., Shi, X. Q., Huang, H., Pillon, N. J., Bilan, P. J., Tu, Y., Klip, A., Ji, R.-R., Zhang, J., ... Mogil, J. S. (2015). Different immune cells mediate mechanical pain hypersensitivity in male and female mice. *Nature Neuroscience, 18*(8), 1081–1083. https://doi.org/10.1038/nn.4053

Starzec-Proserpio, M., Frawley, H., Bø, K., & Morin, M. (2025). Effectiveness of nonpharmacological conservative therapies for chronic pelvic pain in women: A systematic review and meta-analysis. *American Journal of Obstetrics and Gynecology, 232*(1), 42–71. https://doi.org/10.1016/j.ajog.2024.08.006

Stieger, A., Asadauskas, A., Luedi, M. M., & Andereggen, L. (2025). Women's pain management across the lifespan—A narrative review of hormonal, physiological, and psychosocial perspectives. *Journal of Clinical Medicine, 14*(10), 3427. https://doi.org/10.3390/jcm14103427

Stout, N. L., Dierkes, M., Oliveri, J. M., Rockson, S., & Paskett, E. D. (2024). The influence of non-cancer-related risk factors on the development of cancer-related lymphedema: A rapid review. *Medical Oncology, 41*(11), 274. https://doi.org/10.1007/s12032-024-02474-7

Tagaev, T., Anarbaev, A., Kudaiarov, D., Tagaeva, B., Mamatov, S., & Vityala, Y. (2025). Global impact and management in patients with senile osteoporosis: A systematic review. *Asian Journal of Pharmaceutics (AJP), 19*(2). https://doi.org/10.22377/ajp.v19i2.6421

Takahashi, T. A., & Johnson, K. M. (2015). Menopause. *The Medical Clinics of North America, 99*(3), 521–534. https://doi.org/10.1016/j.mcna.2015.01.006

Tank, K., & Rathod, P. (2025). Innovations in pelvic floor rehabilitation: A review of wireless intravaginal biofeedback devices for women. *Health Problems of Civilization.* https://doi.org/10.5114/hpc.2025.152430

Tenório, N., Lima, M. G. A., Leitão, H. A. de S., & Dantas, D. (2025). Mobile applications for assessment and monitoring of breast cancer-related lymphedema: A systematic review. *BioMedInformatics, 5*(4), 62. https://doi.org/10.3390/biomedinformatics5040062

Thuilier, E., Carey, J., Dempsey, M., Dingliana, J., Whelan, B., & Brennan, A. (2024). Virtual rehabilitation for patients with osteoporosis or other musculoskeletal disorders: A systematic review. *Virtual Reality, 28*(2), 93. https://doi.org/10.1007/s10055-024-00980-7

Thurston, K. L., Zhang, S. J., Wilbanks, B. A., Billings, R., & Aroke, E. N. (2023). A systematic review of race, sex, and socioeconomic status differences in postoperative pain and pain management. *Journal of PeriAnesthesia Nursing, 38*(3), 504–515.

Torgbenu, E., Luckett, T., Buhagiar, M. A., Chang, S., & Phillips, J. L. (2020). Prevalence and incidence of cancer related lymphedema in low and middle-income countries: A systematic review and meta-analysis. *BMC Cancer, 20*(1), 604.

UN Affairs, Department of Economic and Social Affairs. (2025). *Disability and development report 2024: Accelerating the realization of the Sustainable Development Goals by, for and with persons with disabilities* (1st ed.). United Nations Research Institute for Social Development.

UNICEF, Department of Economic and Social Affairs. (2023). *Global report on children with developmental disabilities: From the margins to the mainstream.* World Health Organization.

Villegas-Echeverri, J. D., Robert, M., Carrillo, J. F., Green, I., Meinhold-Heerlein, I., Attar, R., Pope, R., & Lamvu, G. (2025). FIGO–IPPS consensus statement: Addressing the global unmet needs of women with chronic pelvic pain. *International Journal of Gynecology & Obstetrics, 169*(3), 1140–1145.

Wagner, B., Steiner, M., Huber, D. F. X., & Crevenna, R. (2022). The effect of biofeedback interventions on pain, overall symptoms, quality of life and physiological parameters in patients with pelvic pain: A systematic review. *Wiener Klinische Wochenschrift, 134*(Suppl. 1), 11–48. https://doi.org/10.1007/s00508-021-01827-w

Wall, T., Feinn, R., Chui, K., & Cheng, M. S. (2015). The effects of the Nintendo[TM] Wii Fit on gait, balance, and quality of life in individuals with incomplete spinal cord injury. *The Journal of Spinal Cord Medicine, 38*(6), 777–783.

Wang, Y., Wei, T., Li, M., Wu, P., Qiang, W., Wang, X., & Shen, A. (2025). Factors influencing the self-management of breast cancer-related lymphedema: A meta-synthesis of qualitative studies. *Cancer Nursing, 48*(5), E339–E353. https://doi.org/10.1097/NCC.0000000000001340

World Health Organization. (2022). *Global report on health equity for persons with disabilities.* World Health Organization. https://www.who.int/teams/noncommunicable-diseases/sensory-functions-disability-and-rehabilitation/global-report-on-health-equity-for-persons-with-disabilities

World Health Organization. (2023). *Disability and health.* World Health Organization. https://www.who.int/news-room/fact-sheets/detail/disability-and-health

Wright, V. J., Schwartzman, J. D., Itinoche, R., & Wittstein, J. (2024). The musculoskeletal syndrome of menopause. *Climacteric: The Journal of the International Menopause Society, 27*(5), 466–472. https://doi.org/10.1080/13697137.2024.2380363

Yamamoto, A., Yamada, E., Ibara, T., Nihey, F., Inai, T., Tsukamoto, K., Waki, T., Yoshii, T., Kobayashi, Y., Nakahara, K., & Fujita, K. (2024). Using in-shoe inertial measurement unit sensors to understand daily-life gait characteristics in patients with distal radius fractures during 6 months of recovery: Cross-sectional study. *JMIR mHealth and uHealth, 12*, e55178. https://doi.org/10.2196/55178

Yosef, A., Allaire, C., Williams, C., Ahmed, A. G., Al-Hussaini, T., Abdellah, M. S., Wong, F., Lisonkova, S., & Yong, P. J. (2016). Multifactorial contributors to the severity of chronic pelvic pain in women. *American Journal of Obstetrics and Gynecology, 215*(6), 760.e1–760.e14. https://doi.org/10.1016/j.ajog.2016.07.023

Zeiaee, A., Zarrin, R. S., Eib, A., Langari, R., & Tafreshi, R. (2021). CLEVERarm: A lightweight and compact exoskeleton for upper-limb rehabilitation. *IEEE Robotics and Automation Letters, 7*(2), 1880–1887.

Chapter 7
Challenges and Future Directions of Emerging Technologies in Disability Rehabilitation

Arshad Nawaz Malik⊙ and Huma Riaz⊙

Abstract This chapter highlights the key challenges, ethical concerns, and future directions in integrating emerging technologies to rehabilitate individuals with physical disabilities. Recent technological innovations have expanded rehabilitation horizons through user-friendly features such as engagement, interaction, precision measurement, concurrent feedback, and real-time monitoring. These technological innovations offer promising solutions to enhance physical function; however, their implementation is often limited, especially in low-resource settings, due to limited accessibility, high costs, training gaps, inadequate infrastructure, and limited cultural adaptability. The chapter sheds light on the ethical considerations. It provides insight into future directions, such as incorporating AI-driven personalised intervention plans to enable flexibility and adaptation in rehabilitation processes, keeping in view the physical, mental, and social aspects of patients. It focuses on the importance of involving end-users in technology development and ensuring inclusivity across age, ability, and socioeconomic backgrounds. These considerations are important to ensure the engagement of all stakeholders in the process, ensuring that technology-driven rehabilitation remains ethical, effective, and sustainable and responsive to the needs of people with physical disabilities.

Keywords Accessibility · Cost-effective · Hybrid care · Patient engagement · Personalised rehabilitation

A. N. Malik (✉) · H. Riaz
Faculty of Rehabilitation and Allied Health Sciences, Riphah International University, Islamabad, Pakistan
e-mail: arshad.nawaz@riphah.edu.pk

H. Riaz
e-mail: huma.riaz@riphah.edu.pk

© The Author(s), under exclusive license to Springer Nature Singapore Pte Ltd. 2026   153
A. N. Malik et al., *Emerging Technologies in the Rehabilitation of Physical Disabilities*,
SpringerBriefs in Modern Perspectives on Disability Research,
https://doi.org/10.1007/978-981-92-1343-6_7

7.1 Introduction

Technologically assisted rehabilitation has transformed the practice landscape through innovation. As discussed in the previous chapters, emerging technologies for managing physical disabilities include Robotics, Virtual reality (VR), Augmented Reality (AR), brain–computer interfaces (BCI), Wearable Sensors, mobile apps, and e-rehabilitation platforms. These emerging technologies aim to reduce the physical disabilities of patients affected by diverse conditions, including neurological, geriatric, women's health, paediatric, and musculoskeletal conditions. Conventional rehabilitation of people with physical disabilities mainly relies on clinical settings, is therapist-dependent, has limited resources, and is geographically restricted. The accessibility of interventions depends on infrastructure, connectivity, digital literacy, awareness, and affordability in adopting emerging technologies. These limitations of conventional approaches include the lack of service delivery accessibility in remote areas, especially in lower-middle-income countries (LMICs). The integration of technology-based rehabilitation has addressed these longstanding issues and provides accessible care, objective-oriented training, monitoring, engaging, and home-based programs. Emerging technologies are promising interventions to improve mobility, activity levels, independence, functional recovery, and community participation among people with physical disabilities. Despite rapid technological advances, certain limitations persist that hinder the effective implementation of interventions. The next section of the chapter critically analyses the significant challenges and future direction of emerging technologies. This chapter will provide a comprehensive framework for policymakers, researchers, and clinicians to prepare a sustainable innovation in the rehabilitation of people with physical disabilities.

7.2 Key Challenges in Technology-Driven Rehabilitation

Despite the transformative potential of emerging technologies in disability rehabilitation, several challenges continue to influence their adoption and implementation. To provide a structured understanding, the key challenges are categorised into the following domains:

7.2.1 Technological and Infrastructure Challenges

Technology-assisted rehabilitation addresses the longstanding issue of access to intervention in remote areas. However, the equity of access remains a challenge, as unequal provision of technological equipment, internet access, digital education, and awareness among different socioeconomic groups in LMICs. Organisational support significantly impacts the adoption of technology-assisted rehabilitation programs in clinical settings. Providing infrastructure, space, financial support, and sufficient staff is crucial to implementing these technology-based interventions. The unavailability or delay in any of these components directly affects the delivery of appropriate training to patients and, overall, reduces the efficacy of the rehabilitation program (Banyai & Brişan, 2024; Liu et al., 2022a).

Few technological devices, including robotics and BCI, offer limited flexibility and portability, as well as limited transferability of training from a controlled environment to the real world. Portable devices are not durable and may be damaged when used vigorously during physical activities. The damage can disrupt data collection and interpretation. There are no guidelines for developing VR platforms, as developers use different interfaces, leading to troubleshooting and technical issues. Few VR software developments require high-performance hardware and software that demand significant data storage and power consumption. There are issues related to the repair or maintenance of equipment, such as in clinical settings, where limited biomedical engineering staff and technical expertise are available to address faults associated with emerging technologies (Garcia-Gonzalez et al., 2022; Lobo et al., 2024).

7.2.2 Human and Skill-Related Challenges

Proper training, confidence, and digital literacy are important aspects for effective implementation. Inadequate skill training, limited staff, lack of proper awareness of technologies, and resistance to change in traditional training lead to poor compliance, unfamiliarity, and reduced efficacy. Emerging technologies require continuous support from information technology personnel to operate and manage the frequent technical glitches. Technological assistance training is not recommended for certain conditions, such as cognitive impairment, motion sickness, vertigo, and a history of fits, and there is no alternative technology-related support for these conditions. The majority of emerging technological devices have not been strongly recommended in clinical practice guidelines for broad, uniform implementation.

The VR headset can cause discomfort, visual strain, headaches, and neck and shoulder strain due to its heavy load and the physical demands it places on the neck and shoulder region. Latency issues are also crucial, as a slight delay in feedback can lead to disorientation and discomfort. VR-induced cybersickness is also a major issue, characterised by nausea, headache, and discomfort. This is because of the sensory conflict theory (Hamad & Jia, 2022). There is a huge gap between the collaboration of healthcare professionals with information technology and engineering, and all are working in isolation. This segregated work leads to ineffective patient recovery outcomes and, eventually, low-quality product development. The current protocol of implementation of technology-based training is "one size fits all", which is not suitable for age diversity, cognitive level, severity of disease, and motivational level (Mehmood et al., 2025).

7.2.3 Economic, Social, and Cultural Challenges

The initial cost of these technological devices is high, and there are no structured reimbursement policies or insurance coverage, especially in LMICs. These challenges put a financial burden on families and patients to manage the cost of equipment. Due to the high cost, it is not feasible to provide uniform, standardised care across the board. Cultural values and norms are also integral parts of the implementation of technologically assisted rehabilitation. A preference for face-to-face intervention hinders the adaptation of the digital platform for home-based settings. Cultural sensitivity and contextual adaptation are important for the effective implementation of technology-oriented programs. Gamified platforms and other digital media should consider cultural values when creating motivational and engaging activities to maximise compliance (Archer & Ellis, 2024).

7.2.4 Ethical, Legal, and Data Challenges

Emerging technologies enhance patients' motivation and active engagement in training programs. However, there are also concerns about patients' autonomy and respect while interacting with these technologies. The regular auto-generated reminders should respect patients' autonomy and privacy to avoid coercive digital push. Multiple digital devices continuously collect data and transfer it to central data centres for monitoring progress and informing future decisions about intervention. However, there is an ethical concern about the data privacy and security of these data centres in the current era, as well as a lack of robust data protection and privacy policies, especially in LMICs (Kirsch & Fluet, 2024).

An ethical concern has been raised about whether advanced technologies are accessible to everyone or if only a few will benefit from them, given their high cost. In addition, who will benefit more, the companies, healthcare professionals, caregivers, or patients? The basic requirement of the internet is also a key factor regarding the provision of telerehabilitation. A greater focus on digital platforms for intervention also increases children's screen time. There are also certain problems related to children's physical development due to insufficient physical activity.

There are also no uniform platforms or standardised guidelines for the proper implementation of technology-assisted rehabilitation, and no policy or regulation for the adaptation of digital technologies in health care, especially in LMICs. A limited dataset can lead to errors or overfitting in predictive models for large language models in machine learning and AI-driven solutions. There is limited clinical validation of emerging technologies and a lack of long-term effectiveness data for effective implementation (Tuah et al., 2021a).

7.3 Future Directions in Rehabilitation Technologies

Innovative and emerging technologies have led to a paradigm shift from therapist-dependent to autonomous care delivery. However, there are still areas to improve in future to provide comprehensive, personalised. Affordable and ethical practice through new advancements. The following are the critical areas for future development and practice.

7.3.1 AI-Driven Personalised Rehabilitation

An AI algorithm will decide the future rehabilitation services with accuracy and precision. Early, accurate diagnosis; prognosis; screening; factor identification; prediction; and guidance on future planning will change the dynamics of practice. The future perspective includes an AI-driven, personalised plan based on complete individual data and the inclusion of all physical, mental, and social aspects of patients, which supports better clinical decisions. Digital markers will play a significant role in early detection and enable the therapist to make decisions based on AI-supported data.

Multiple technologies have not been incorporated into routine clinical settings or made accessible in low-income countries. There is a dire need to provide cost-effective, accessible technologies to improve patient outcomes. Gamification is a common approach in physical rehabilitation to achieve functional status through an interactive, engaging platform. All games have been designed for patients; however, caregivers are also an important part of the whole rehabilitation process. Engagement of all stakeholders in the process will enhance recovery. The future direction includes a personalised, individualised intervention plan tailored to each patient's context. Such changes provide flexibility and adaptation of rehab processes. Technology is highly dynamic and innovative, and it rapidly changes with new versions; this is also one of the challenges of adopting such robust changes, both for healthcare professionals and patients. It is also important to develop such innovative technology-based equipment, as it needs expertise (Tuah et al., 2021b).

It is important to align with and interact with AI and machine learning to provide accurate, precise data and deliver personalised training programs. The combination of robotics integration with VR and wearable sensors provides an enriching and engaging training environment (Liu et al., 2022). Multiple sensors can simultaneously provide comprehensive data and enable functional evaluation. The decentralisation of rehabilitation is an important future goal, and the incorporation of sensors into daily activity equipment to monitor activities without disturbing the patient's routine. Complete autonomous rehabilitation plans, from assessment through at-home rehabilitation planning, should be devised to achieve significant outcomes. The combination of EMG, eye tracking, sensors, and an AI-driven algorithm with personalisation and real-time feedback can improve the efficacy of emerging technologies (Latif et al., 2024; Wei & Wu, 2023).

7.3.2 User-Centred and Inclusive Design

The future design of technologies should be flexible, portable, adjustable, multi-modal, and compatible with other assistive technologies for proper alignment and operation. The innovative design should be feasible and suitable for all paediatric to geriatric patients, from non-communicative to communicative, covering a large population. Technologies have engagement features that provide an interactive environment; however, they are more engaging through challenges, rewards, real-time feedback, and family and peer involvement through gamification. The inclusion of specific, personalised, and adaptive training, incorporating AI and machine learning, for clinical staff should be added to create a more conducive environment.

The use of a brain–computer interface still has a few challenges in its proper and full-scale utilisation due to precision issues. The involvement of nonmotor activities also hinders the utilisation of BCI. Further studies using advanced neuroimaging can be beneficial for developing a user-friendly system to enhance recovery (Simon et al., 2021). Communication challenge with assistive technologies is also a key factor, as patients with physical disabilities, especially children, may have cognitive impairment as well. They cannot understand the commands and sequence of the technology-oriented process, thereby delaying rehabilitation. Innovative technologies that operate with minimal command gestures are required to address this barrier (Dhar et al., 2023; Kouijzer et al., 2023).

7.3.3 Low-Cost and Scalable Solutions

Technological-assisted rehabilitation should be equally provided to the entire population, regardless of their socioeconomic status. Future rehabilitation technologies should be cost-effective, freely available, and amenable to local manufacturing to enable their implementation in LMICs. Digital, innovative, and simple platforms using mobile apps and wearable technologies should be developed to enable home-based remote monitoring and real-time feedback. The inclusion of these devices enhances adherence, long-term follow-up, and reduces travel costs and burdens on healthcare systems. The future intervention should promote community integration, care for chronic diseases, and enhance social support (Gower et al., 2024).

7.3.4 Interdisciplinary and Collaborative Approaches

The future belongs to strong interdisciplinary collaboration for effective, beneficial outcomes in the healthcare system. Stakeholders include therapists, patients, software engineers, computer scientists, biomedical engineers, and family members who should be involved in the process. Reflections of the real world in activities enhance patient compliance and improve participation levels. The engagement of all stakeholders ensures adherence, sustainability, and acceptability in the proper utilisation of emerging technologies. CPDs should be integral to all clinical staff, patients, and family members to improve digital literacy. Digital education provides an understanding of the process, basic AI concepts, and data interpretation to make decisions about interventions (Mirinchev, 2024).

7.3.5 Ethical and Regulatory Considerations

In the future, it is important to address key ethical issues, including human privacy, data security, cybersecurity, bias in AI algorithms, and equity across all groups. Ethical concerns associated with AI should be explored, and work is required on fairness, accountability, and judicious utilisation. Further research into social and cultural beliefs and values to inform better future adaptive strategies in emerging technologies. Uniform framework and policies should be devised and approved by all regulatory and official bodies, with a complete description of SOPs for standardised practices. Validated outcomes should be designed to evaluate the effectiveness of different populations. Evidence-based practice guidelines with recommendations on the role of emerging technologies in rehabilitation should be prepared to guide healthcare professionals (Mehmood et al., 2025).

7.4 Integrating Emerging Technologies into Sustainable Rehabilitation Models

WHO (2019) recommendations on digital interventions for health system strengthening emphasise that technology-based and digital health interventions can support and strengthen the health system, including rehabilitation, by aligning digital tools with global health coverage strategies, improving workforce capacity, and embedding digital outcomes reporting in health information systems (World Health Organization, 2019). A systems-level approach is essential to establish technology-integrated sustainable rehabilitation models. These models should adapt environmentally accountable service re-designs. Adopting a triple bottom line framework can serve this purpose by balancing rehabilitation across the strategic pillars of clinical effectiveness, economic efficiency, and environmental responsibility (Palstam et al., 2022). To explain, the disability management models should have demonstrated clinical efficacy in improving functional outcomes, social participation, and quality of life. The technology-enabled models should have cost-utility and service scalability, as well as efficient resource utilisation, to reduce unnecessary travel, material consumption, and energy use at facility sites.

7.4.1 Combining Conventional Therapy with Tech-Driven Approaches

In hybrid care rehabilitation pathways, sustainability is optimised when technology-led interventions are offered alongside conventional treatment for individuals with disabilities. This model typically includes an initial assessment to assess diagnostic accuracy, comprehensive clinical evaluation and reasoning, followed by technology-enabled monitoring and therapeutic exercise delivery and progression. The final reassessment is conducted either face-to-face or remotely, based on the predefined clinical indicators.

There are some key determinants for sustainable rehabilitation models. The convenience and accessibility achieved in remote interventions identify factors such as reduced travel cost and time, reduced caregiver burden, and easier scheduling, contributing to environmental sustainability and economic efficiency. Furthermore, optimisation of workforce utilisation and burden, reduction of unnecessary facility usage, and enhanced rural access can be achieved through hybridisation. In general, the hybrid models have received broad support, as they maintain clinical supervision and rigour through periodic face-to-face visits, allow individualised intervention based on the phase of recovery, and preserve convenience through remote follow-ups.

A cross-sectional study was conducted on individuals with diverse physical and neurological disabilities to evaluate their preferences for rehabilitation service delivery modality. The key findings inform us that heterogeneity is central; individuals with disabilities did not demonstrate a similar preference for a single modality. Some preferred in-person services, such as hands-on therapy, others liked the telehealth option for convenience and flexibility, whereas a significant proportion favoured hybrid approaches. The implications guide us that sustainable rehabilitation systems should be flexible and adaptable (Anderson et al., 2022).

7.4.2 Strategies for Long-Term Adoption, Adherence, and Engagement

Long-term patient engagement in technology-embedded, digitally driven rehabilitation models is identified as a major barrier, as only short-term benefits have been studied and reported to date. Without tailored strategies, adherence to intervention protocols wanes over time. A multidisciplinary strategy is needed to achieve long-term adoption, adherence, and engagement with technology-integrated disability management. Below are the evidence-based strategies grounded in behavioural theory, implementation science, patient-centred care design, and health system integration (Al Mahmud et al., 2026).

- Integrate behavioural change techniques in technology-assisted rehabilitation models: Behavioural science models, such as the Health Belief Model or the

Technology Acceptance Model, explain the contributing factors, including ease of use, perceived usefulness, and self-efficacy, for sustained engagement. Several motivational features, such as goal-setting for interventions, automated alerts and reminders, gamification elements (levels, rewards, and badges), and peer support, can improve adherence to rehabilitation for chronic conditions (Lu et al., 2025).

- The hybrid rehabilitation model of care maintains a better therapeutic alliance than digital health interventions delivered alone. Some strategies include initial on-site evaluations to assess diagnostic accuracy, regular clinicians' check-ins or messaging via web portals or apps, sharing progress summaries between therapists and patients, therapists reviewing wearable or patient inputs in asynchronous mode, and scheduled remote appointments to follow patient progress (Lang et al., 2022).

- The objective measurement of adherence monitoring outcomes or standardised adherence metrics (for example, % of prescribed sessions completed) can help and allow early identification of disengagement, which can trigger adaptive support. Practices include using wearable sensor data to track engagement and activity objectively, and enabling built-in reporting functions to visualise trends and intervene promptly (Lang et al., 2022).

- E-health or digital literacy is the key mediator for technology-enabled rehabilitation models. People with more digital skills have shown more interest and retention in the protocol. Approaches such as providing instructional support (for those with limited technology exposure), pre-intervention assessments of technology usage, apprehension, or literacy, navigation features in app tutorials, and simplified interfaces for diverse functional levels can significantly enhance engagement and confidence in using digital rehab systems (Georgas et al., 2025).

- Patient preference for the rehab model selection is of significant importance. Engaging all stakeholders, such as patients, clinicians, and sometimes caregivers, in developing co-design protocols can align interventions to their real needs and preferences. Introducing customised features to adjust to disability type, severity, and daily routines can reduce barriers to adoption and enhance ownership of the rehab process (Al Mahmud et al., 2026).

- The technical and contextual barriers should be identified and considered while developing disability management plan. Sustained use can be compromised by poor connectivity, device inaccessibility, and workflow fit. Strategies such as low-bandwidth designs, offline support, minimal user burden, and automated clinician alerts can preempt these contextual obstacles (Al Mahmud et al., 2026).

Together, these strategies can contribute to robust, technology-integrated disability management models that enable sustained recovery and functional engagement.

7.5   Conclusion and Future Direction

Technology-integrated rehabilitation represents a fundamental paradigm shift in disability rehabilitation. By combining human expertise with technological innovation, modern rehabilitation systems can become more precise, personalised, accessible, and effective. However, their transformative potential can only be realised if critical challenges are addressed systematically. The applicability and feasibility of technology-driven rehabilitation depend on technological limitations, infrastructure gaps, skill deficits, socioeconomic barriers, and ethical concerns. If these issues are not carefully addressed, technological innovation may unintentionally deepen existing inequalities rather than help reduce them.

The future of rehabilitation lies in intelligent, adaptive, and user-centred systems that complement, not replace, clinical expertise. Artificial intelligence-driven personalisation, predictive analytics, and wearable monitoring systems hold promise for optimising functional outcomes and long-term engagement. However, technological progress must be guided by inclusivity, respect for cultural differences, and design approaches that accommodate diverse needs. This is essential to ensure that people of different ages, disability types, and socioeconomic backgrounds can equitably benefit from rehabilitation technologies.

Sustainable progress will depend on interdisciplinary collaboration among clinicians, engineers, researchers, policymakers, and community stakeholders. Investment in providing training, scalable low-cost solutions for resource-constrained settings, and clear ethical and regulatory frameworks will be essential to support responsible innovation. Furthermore, continuous evaluation through robust clinical metrics, patient-reported outcomes, and cost-effectiveness analyses will determine the true value of these technologies within rehabilitation systems. Ultimately, the integration of emerging technologies into rehabilitation is not merely a technical endeavour but a human-centred one. The goal is not innovation for its own sake, but meaningful enhancement of function, participation, independence, and quality of life for individuals with disabilities. Achieving this vision requires sustained research, adaptive implementation models, and a shared commitment to equity, ethics, and long-term sustainability.

References

Al Mahmud, A., Joachim, S., Jayaraman, P. P., Learmonth, C., Tyagi, S., Forkan, A. R. M., Shuakat, M., Wickramasinghe, N., Wheeler, J., Best, S., & Trainer, A. (2026). Digital Health Interventions to Support Chronic Disease Management: Systematic Scoping Review. *JMIR mHealth and uHealth, 14*, Article e63742. https://doi.org/10.2196/63742

Anderson, R., Thompson, N., Hanson, M., Kurzweil, E., Metzger, K., & Johnstone, B. (2022). Preferences for In-person, Telehealth, or Hybrid Rehabilitation Services for Individuals with Diverse Disabilities. *Archives of Physical Medicine and Rehabilitation, 103*(12), Article e69. https://doi.org/10.1016/j.apmr.2022.08.606

Archer, K. R., & Ellis, T. D. (2024). Advances in rehabilitation technology to transform health. *Physical Therapy, 104*(2), pzae008.

Banyai, A. D., & Brișan, C. (2024). *Robotics in Physical Rehabilitation: Systematic Review., 12*(17), 1720.

Dhar, E., Upadhyay, U., Huang, Y., Uddin, M., Manias, G., Kyriazis, D., Wajid, U., AlShawaf, H., & Syed Abdul, S. (2023). A scoping review to assess the effects of virtual reality in medical education and clinical care. *Digital Health, 9,* 20552076231158024.

Garcia-Gonzalez, A., Fuentes-Aguilar, R. Q., Salgado, I., & Chairez, I. (2022). A review on the application of autonomous and intelligent robotic devices in medical rehabilitation. *Journal of the Brazilian Society of Mechanical Sciences and Engineering, 44*(9), 393.

Georgas, K., Bromis, K., Vagenas, T. P., Giannakopoulou, O., Vasileiou, N., Kouris, I., Haritou, M., & Matsopoulos, G. K. (2025). eHealth literacy assessment as a promoter of user adherence in using digital health systems and services. A case study for balance physiotherapy in the TeleRehaB DSS project. *Frontiers in Digital Health, 7.* https://doi.org/10.3389/fdgth.2025.153 5582

Gower, V., Aprile, I., Falchini, F., Fasano, A., Germanotta, M., Randazzo, M., Spinelli, F., Trieste, L., Gramatica, F., & Turchetti, G. (2024). Cost analysis of technological vs. Conventional upper limb rehabilitation for patients with neurological disorders: An Italian real-world data case study. *Frontiers in Public Health, 12,* 1445099.

Hamad, A., & Jia, B. (2022). How virtual reality technology has changed our lives: An overview of the current and potential applications and limitations. *International Journal of Environmental Research and Public Health, 19*(18), 11278.

Kirsch, N., & Fluet, G. (2024). Ethical and Moral Considerations: Telerehabilitation. In *Emerging Technologies in Healthcare* (pp. 79–95). CRC Press.

Kouijzer, M. M., Kip, H., Bouman, Y. H., & Kelders, S. M. (2023). Implementation of virtual reality in healthcare: A scoping review on the implementation process of virtual reality in various healthcare settings. *Implementation Science Communications, 4*(1), 67.

Lang, S., McLelland, C., MacDonald, D., & Hamilton, D. F. (2022). Do digital interventions increase adherence to home exercise rehabilitation? A systematic review of randomised controlled trials. *Archives of Physiotherapy, 12*(1), 24. https://doi.org/10.1186/s40945-022-00148-z

Latif, A., Al Janabi, H. F., Joshi, M., Fusari, G., Shepherd, L., Darzi, A., & Leff, D. R. (2024). Use of commercially available wearable devices for physical rehabilitation in healthcare: A systematic review. *British Medical Journal Open, 14*(11), Article e084086.

Liu, C., Lu, J., Yang, H., & Guo, K. (2022). Current state of robotics in hand rehabilitation after stroke: A systematic review. *Applied Sciences, 12*(9), 4540.

Lobo, P., Morais, P., Murray, P., & Vilaça, J. L. (2024). Trends and innovations in wearable technology for motor rehabilitation, prediction, and monitoring: A comprehensive review. *Sensors, 24*(24), 7973.

Lu, T., Lin, Q., Yu, B., & Hu, J. (2025). A systematic review of strategies in digital technologies for motivating adherence to chronic illness self-care. *Npj Health Systems, 2*(1), 13. https://doi.org/10.1038/s44401-025-00017-4

Mehmood, F., Mumtaz, N., & Mehmood, A. (2025). *Next-Generation Tools for Patient Care and Rehabilitation: A Review of Modern Innovations., 14*(3), 133.

Mirinchev, R. (2024). *Multidisciplinary Approach in the Modern Rehabilitation Industry.* 179–188.

Palstam, A., Sehdev, S., Barna, S., Andersson, M., & Liebenberg, N. (2022). Sustainability in physiotherapy and rehabilitation. *Orthopaedics and Trauma, 36*(5), 279–283.

Simon, C., Bolton, D. A. E., Kennedy, N. C., Soekadar, S. R., & Ruddy, K. L. (2021). Challenges and Opportunities for the Future of Brain-Computer Interface in Neurorehabilitation. *Frontiers in Neuroscience, 15.* https://doi.org/10.3389/fnins.2021.699428

Tuah, N. M., Ahmedy, F., Gani, A., & Yong, L. N. (2021a). A survey on gamification for health rehabilitation care: Applications, opportunities, and open challenges. *Information, 12*(2), 91.

Tuah, N. M., Ahmedy, F., Gani, A., & Yong, L. N. (2021b). A Survey on Gamification for Health Rehabilitation Care: Applications, Opportunities, and Open Challenges. *Information, 12*(2), 91. https://doi.org/10.3390/info12020091

Wei, S., & Wu, Z. (2023). The application of wearable sensors and machine learning algorithms in rehabilitation training: A systematic review. *Sensors, 23*(18), 7667.

World Health Organization. (2019). *Recommendations on digital interventions for health system strengthening – Evidence and recommendations* (WHO-RHR-19.10). World Health Organization. https://www.who.int/publications/i/item/WHO-RHR-19.10